Corrosion of Reinforcement in Concrete — Monitoring, Prevention and Rehabilitation

Papers from EUROCORR '97

Trondheim, Norway, 1997

European Federation of Corrosion Publications
NUMBER 25

Corrosion of Reinforcement in Concrete — Monitoring, Prevention and Rehabilitation

Papers from EUROCORR '97
Trondheim, Norway, 1997

Edited by
J. MIETZ, B. ELSENER AND R. POLDER

Published for the European Federation of Corrosion by The Institute of Materials

CRC Press
Taylor & Francis Group
Boca Raton London New York

CRC Press is an imprint of the
Taylor & Francis Group, an **informa** business

First published 1998 by IOM Communications Ltd

Published 2019 by CRC Press
Taylor & Francis Group
6000 Broken Sound Parkway NW, Suite 300
Boca Raton, FL 33487-2742

© 1998 The Institute of Materials
CRC Press is an imprint of Taylor & Francis Group, an Informa business

First issued in paperback 2019

No claim to original U.S. Government works

ISBN 13: 978-0-367-44768-7 (pbk)
ISBN 13: 978-1-86125-083-4 (hbk)

**Visit the Taylor & Francis Web site at
http://www.taylorandfrancis.com**

**and the CRC Press Web site at
http://www.crcpress.com**

Neither the EFC nor The Institute of Materials
is responsible for any views expressed
in this publication

Design and production by
SPIRES Design Partnership

Contents

Part 5 – Cathodic Protection   159

European Federation of Corrosion Publications
Series Introduction

The EFC, incorporated in Belgium, was founded in 1955 with the purpose of promoting European co-operation in the fields of research into corrosion and corrosion prevention.

Membership is based upon participation by corrosion societies and committees in technical Working Parties. Member societies appoint delegates to Working Parties, whose membership is expanded by personal corresponding membership.

The activities of the Working Parties cover corrosion topics associated with inhibition, education, reinforcement in concrete, microbial effects, hot gases and combustion products, environment sensitive fracture, marine environments, surface science, physico–chemical methods of measurement, the nuclear industry, computer based information systems, the oil and gas industry, the petrochemical industry and coatings. Working Parties on other topics are established as required.

The Working Parties function in various ways, e.g. by preparing reports, organising symposia, conducting intensive courses and producing instructional material, including films. The activities of the Working Parties are co-ordinated, through a Science and Technology Advisory Committee, by the Scientific Secretary.

The administration of the EFC is handled by three Secretariats: DECHEMA e. V. in Germany, the Société de Chimie Industrielle in France, and The Institute of Materials in the United Kingdom. These three Secretariats meet at the Board of Administrators of the EFC. There is an annual General Assembly at which delegates from all member societies meet to determine and approve EFC policy. News of EFC activities, forthcoming conferences, courses etc. is published in a range of accredited corrosion and certain other journals throughout Europe. More detailed descriptions of activities are given in a Newsletter prepared by the Scientific Secretary.

The output of the EFC takes various forms. Papers on particular topics, for example, reviews or results of experimental work, may be published in scientific and technical journals in one or more countries in Europe. Conference proceedings are often published by the organisation responsible for the conference.

In 1987 the, then, Institute of Metals was appointed as the official EFC publisher. Although the arrangement is non-exclusive and other routes for publication are still available, it is expected that the Working Parties of the EFC will use The Institute of Materials for publication of reports, proceedings etc. wherever possible.

The name of The Institute of Metals was changed to The Institute of Materials with effect from 1 January 1992.

A. D. Mercer
EFC Series Editor,
The Institute of Materials, London, UK

EFC Secretariats are located at:

Dr B A Rickinson
European Federation of Corrosion, The Institute of Materials, 1 Carlton House Terrace, London, SW1Y 5DB, UK

Mr P Berge
Fédération Européene de la Corrosion, Société de Chimie Industrielle, 28 rue Saint-Dominique, F-75007 Paris, FRANCE

Professor Dr G Kreysa
Europäische Föderation Korrosion, DECHEMA e. V., Theodor-Heuss-Allee 25, D-60486, Frankfurt, GERMANY

Preface

The durability of reinforced concrete, one of the most widely used construction materials, is a topic of major importance throughout the world with large economic implications. In Europe, where a great part of the infrastructure is already built, the rehabilitation of reinforced concrete is becoming the most important point in the maintenance and life-cycle performance of structures. Thus, interest of engineers and owners of structures in non-destructive monitoring techniques, in repair methods without concrete removal and in new preventive measures is increasing and industry is offering a large variety of continuously improved products to meet those needs.

The present volume of the EFC series compiles the full papers presented in the successful session "Corrosion of Steel in Concrete" at EUROCORR '97, held at Trondheim, Norway. The topics highlight the main current areas of technical development in this rapidly evolving field, including monitoring of steel reinforcement corrosion, prevention of corrosion and electrochemical repair methods. Seventeen papers were accepted for publication after peer review, including papers from Denmark, Germany, Italy, The Netherlands, Norway, Spain, Switzerland and the United Kingdom. The papers were grouped under five major headings:

- Corrosion mechanisms,
- Reference electrodes, supplementary protection measures,
- Corrosion rate measurement,
- Electrochemical realkalisation and chloride removal,
- Cathodic protection.

We thank all authors who have shared their valuable experience with others and stimulated discussions. The editors encourage the readers to evaluate for themselves conclusions based upon the results given in the papers and the included references.

J. MIETZ, B. ELSENER AND R. POLDER
Editors

Part 1

Corrosion Mechanisms

1

Effect of Potential Drop in Migration Testing of Chloride Diffusivity in Concrete

T. ZHANG and O. E. GJØRV

Department of Building Materials, Norwegian University of Science and Technology (NTNU),
N-7034 Trondheim, Norway

ABSTRACT

In the calculation of chloride diffusivity based on migration testing, the external potential is normally applied without any correction for the potential drop in the system outside the test specimen. In the present paper, it is shown that this potential drop may give an error of up to 20% of the chloride diffusivity obtained based on the nominal (external) applied potential. For different types of concrete (w/c = 0.40–0.60) and different salt solutions (1–3% NaCl in the upstream cell and 0.3M NaOH in the down-stream cell) and different levels of applied voltage (10–60 V), a total potential drop in the migration system varying from approximately 2.5 to 3.0 V was observed. Since measurements of such a potential drop make the routine testing of chloride diffusivity more elaborate, a simple correction factor of 2.5 V for the potential drop in the migration system outside the test specimen may be introduced.

1. Introduction

Over recent years, migration testing of chloride diffusivity has increasingly been adopted for accelerated testing not only of the resistance of concrete against chloride penetration but also for reflecting the permeability of very dense cementitious materials [1,2]. In the calculation of the chloride diffusivity, however, the external applied potential is normally used without any correction for the potential drop in the migration system outside the test specimen [3]. Since this potential drop is so high that it should not be ignored [4], a measurement of the real potential difference across the test specimen should always be made. Such a measurement will, however, make the routine testing of chloride diffusivity more elaborate. If the potential drop in the migration system does not vary too much, a more simple correction for the routine testing may be introduced. In order to provide more information about the potential drop in the migration testing, an experimental programme was carried out, the results of which are reported in this paper.

2. Theoretical Background

When an external potential is applied to an electrolytic, aqueous solution system, a potential drop will occur in all interfaces between the solids and the solution due to the combined effect of the electrical double layer and polarisation. The distribution of the potential in an electrolytic cell is schematically shown in Fig. 1, where ΔU is the total external potential difference between the two electrodes, while $\Delta 1$ and $\Delta 2$ are the potential drops at the interface between the electrodes and the electrolytic solution. This type of potential drop, which is mainly determined by the physical-chemical properties of the system, can be denoted as an 'intrinsic potential drop' in the system. Thus, in a system for migration testing of concrete, the following relationship exists:

$$\Delta U_{ext} = \Delta U_{sp} + \Delta U_{dr} \tag{1}$$

where ΔU_{ext} = external applied potential, ΔU_{sp} = potential difference across the specimen, and ΔU_{dr} = total potential drop outside the specimen.

It is the potential difference across the specimen (ΔU_{sp}) that contributes to the migration of chloride ions inside the specimen, which should be used for calculation of the chloride diffusivity. Any potential drop outside the test specimen only reduces the electrical field in the specimen. The total potential drop ΔU_{dr} is composed of two parts, the 'intrinsic potential drop' ΔU_{in} and the 'solution-induced potential drop' ΔU_{sol}:

$$\Delta U_{dr} = \Delta U_{in} + \Delta U_{sol} \tag{2}$$

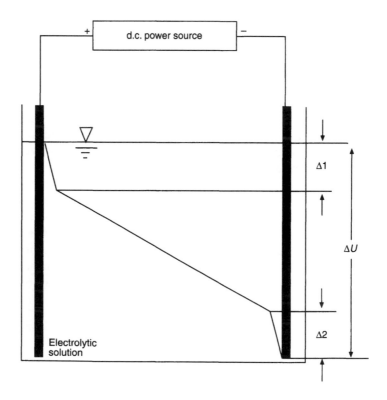

Fig. 1 *Potential distribution in an electrolytic cell [4].*

The intrinsic potential drop ΔU_{in} is determined by the physical–chemical properties of the system. In a system for migration testing of chloride diffusivity in concrete, this potential drop is mainly due to the polarisation of the electrodes induced by the d.c. current and the formation of an electrical double layer both at the interface between the solutions and the specimen and between the solutions and the electrodes. According to the theory of electrical double layer, the formation of an electrical double layer is the result of a selective adsorption of ions from the solution on the electrode surface. Thus, in such a system this potential drop is determined by the properties of the electrodes, the concrete as well as the solution. Since the solutions in the up-stream and down-stream cells usually are different, the potential drops at the cathode and the anode will also be different [5]. The reversible dissociation potential of water can also be considered as being part of the intrinsic potential drop. At pH = 13 and 25°C, this potential makes up a value of 0.463 V.

The solution-induced potential drop ΔU_{sol} mainly comes from the ohmic potential loss ΔU_{ohm} in the solutions, which depends on the electrical conductivity of the solutions, the geometrical parameters of the experimental set-up and the current intensity:

$$\Delta U_{ohm} = I \cdot R_{sol} = \frac{\Delta U_{ext}}{R_{tot}} \cdot R_{sol} = \frac{\Delta U_{ext}}{\dfrac{R_{sp}}{R_{sol}} + 1} \tag{3}$$

where ΔU_{ohm} is the ohmic potential drop, ΔU_{ext} is the external potential, I is the current intensity, R_{sol} is the resistance of the solutions, R_{sp} is the resistance of the test specimen and R_{tot} is the total resistance between the two electrodes ($R_{sol} + R_{sp}$).

From eqn (3) it can be seen that ΔU_{ohm} depends on the ratio of the resistance of the specimen to that of the solutions (R_{sp}/R_{sol}) and the external applied potential U_{ext}. The electrical resistivity of a concrete specimen is usually very high compared to that of the salt solutions used. The electrical resistivity of a concentrated salt solution (≥ 0.001 M) is usually less than 1 Ωm, while the resistivity of a saturated concrete may vary from 50 to 100 Ωm. For a very dense concrete, the resistivity may even reach up to 500 Ωm. Thus, the ratio R_{sp}/R_{sol} may be so high that this type of potential drop becomes negligible for the external potentials normally applied.

3. Experimental
3.1. Materials and Specimen Preparation

Based on an ordinary Portland cement, five different types of concrete were produced with w/c ratio and compressive strength varying from 0.40 to 0.60 and from 41.4 to 92.3 MPa, respectively (Table 1). From each concrete mixture, a number of 200 × 100 dia. cylinders were produced and stored in water at 20 ± 2°C for at least three months. Then, each cylinder was slightly surface dried and embedded in a 5-mm thick epoxy mortar before being cut into 50 mm thick slices and subjected to a standard procedure for water saturation [3].

Table 1. Concrete mixtures

| Mix No. | w/c | Mix proportions (kgm⁻³) | | | | | Compressive strength (MPa) |
		Cement	Water	Silica fume	Aggregate 0–8 mm	8–16 mm	
A	0.40	418	167	–	1101	878	68.5
B	0.50	387	194	–	1020	814	51.7
C	0.60	333	200	–	1024	864	41.4
SFA	0.40	357	159	40	995	836	92.3
SFB	0.50	342	190	38	950	816	74.3

3.2. Test Procedures

An experimental setup for steady state migration testing of chloride diffusivity was used as described elsewhere [3,6], and two different procedures for measurement of the potential difference across the test specimen (ΔU_{sp}) were used (Figs 2 and 3). The thickness of the concrete specimens was 50 mm.

In Test Procedure 1, two small probes were tightly attached to the surface of both sides of the specimen (Fig. 2), while in Test Procedure 2, two capillaries filled with saturated KCl solution and two standard reference electrodes (Calomel commercial type K401) were connected through salt bridges (Fig. 3). To obtain comparable results, both procedures were successively applied to the same test specimens.

In order to investigate the effect of the solution-induced potential drop, two source chloride concentrations (1 and 3% NaCl) were used in the upstream cell, while in the downstream cell, a 0.3M NaOH solution was kept constant. In addition, two different positions of the mesh electrodes were also used. In position A, the electrodes were placed as close as possible to the test specimen, while in position B, the mesh electrodes were placed approximately 25 mm away from each side of the specimen. A d.c. power source with an output range of 0–60 V was applied.

4. Results and Discussion

Some selected experimental data are plotted in Figs 4(a) and (b), from which it can be seen that there is a good linearity between the external potential U_{app} and the potential difference across the specimen U_{sp}. However, the observed potential difference across the specimen (U_{sp}) by use of Test Procedure 2 was smaller than that of Test Procedure 1. This may be due to a tighter attachment of the probes to the specimen surface in Procedure 1 than it was possible to obtain with the capillaries used in Procedure 2. It was more difficult to control an accurate position of the capillaries, the tips of which may have been outside the range of the electrical double

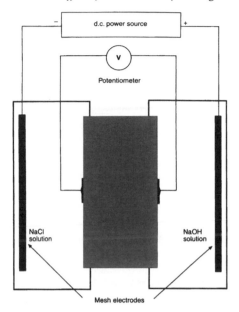

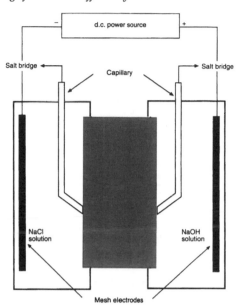

Fig. 2 *Experimental set-up for Test Procedure 1.*

Fig. 3 *Experimental set-up for Test Procedure 2.*

layer on the surface of the specimen. Hence, the measured U_{sp} in Test Procedure 2 may include the potential drops in the electrical double layer. Therefore, the measurements based on Test Procedure 1 should be more correct, and thus the following discussion is based on the results obtained by use of this test procedure.

When the external potential increased from 10 to 60 V, the total potential drop outside the test specimen (U_{dr}) increased from approximately 2.5 to 3.0 V for all types of concrete investigated.

In Table 2, a comparison of the total potential drops obtained by use of the two test procedures and that obtained by McGrath and Hooton [5] is given. In the latter, an OPC concrete with a w/c ratio of 0.49 and a chloride source solution of 0.5M NaCl + 0.3M NaOH were used, while the potential difference across the concrete specimens was measured by use of glass Luggin capillaries and Ag/AgCl standard electrodes.

The test procedure applied by McGrath and Hooton is similar to Test Procedure 2, and from Table 2 it can be seen that the results they obtained are also similar to those obtained by Test Procedure 2.

All test results obtained from the various concrete mixtures showed very small deviations, which indicate that the potential drop was independent of the different types of concrete investigated.

On comparing the results obtained from the two different chloride concentrations and the two different positions of electrodes it can be seen that the effect of these variables on the potential drop was also negligible.

The test results further showed that the potential drop in the system only varied within a small range (2.5–3.0 V) for external applied potentials varying from 10 to 60 V. This is mainly due to the low electrical resistance of the solutions and the high resistance of the concrete specimens and thus the high R_{sp}/R_{sol} ratio. Hence, the

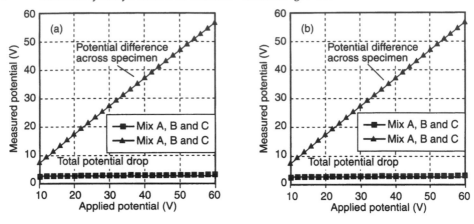

Fig. 4 *Results based on Test Procedure 1, electrode position B and (a) 1% NaCl and (b) 3% NaCl solution.*

intrinsic potential drop makes up most of the total potential drop, while the solution-induced potential drop can be ignored for the practical range of applied potentials.

Based on the above test results, it appears that for routine testing, a simple correction for the potential drop may be introduced. Thus, in order to have a minimum error, a correction factor of 2.5 V appears to be appropriate for the external potential range of 10–60 V:

$$\Delta U_{sp} = \Delta U_{app} - 2.5 \tag{4}$$

where ΔU_{sp} is the potential difference across the specimen, which should be used for calculation of the chloride diffusivity, while ΔU_{app} is the external applied potential between the two electrodes.

By use of eqn (4), the required external potential can easily be determined according to the desired potential across the specimen. Or conversely, the real potential difference across the specimen can be estimated based on the applied potential.

Table 3 shows a comparison between the chloride diffusivities calculated on the basis of the nominal potential (i.e. the external applied potential) U_{app}, the real potential difference across the specimen (i.e. the measured potential) U_{sp} and the estimated potential difference across the specimen U_{est} by using eqn (4) [6,7].

Table 2. *Observed potential drops in the concrete migration system*

Applied potential U_{app} (V)	Total potential $U_{app} - U_{sp}$ (V)		
	Test Procedure 1	Test Procedure 2	Ref. [5]
12	2.52	1.88	2.08
20	2.65	2.03	2.20
30	2.79	2.21	2.36

Table 3. *A comparison of chloride diffusivities calculated with and without corrections*

Mix No.	U_{est}	U_{sp}	U_{app}	Chloride diffusivity (cm² s⁻¹) × 10⁻⁸		
	(V)	(V)	(V)	U_{app}	U_{sp}	U_{est}
A	12	11.96	14.5	0.52	0.63	0.63
	24	23.80	26.5	0.76	0.85	0.84
	36	35.71	38.5	1.22	1.31	1.30
SFA	12	12.06	14.5	0.06	0.07	0.07
	36	35.88	38.5	0.09	0.10	0.10
	60	59.4	62.5	0.10	0.11	0.11
SFB	12	11.96	14.5	0.10	0.12	0.12
	36	35.69	38.5	0.18	0.20	0.20
	60	59.3	62.5	0.23	0.24	0.24

From the data shown in Table 3, it can be seen that an error in the calculation of chloride diffusivity of up to 20% is made only if an external (nominal) potential of 12 V is applied without any correction. The use of Eqn (4), however, shows that the errors are very small for the whole range of external potentials applied.

5. Conclusions

From the experimental investigation carried out, the following conclusions appear to be warranted:

1. In the concrete migration system, a potential drop outside the test specimen exists, which gave an error of up to 20% of the chloride diffusivity obtained on the basis of the nominal (external) potential applied.

2. For the types of concrete investigated (w/c = 0.40–0.60) and for the chloride salt concentrations applied (1–3% NaCl) as well as for the different positions of the electrodes (close or remote from the test specimen), the total potential drop in the migration system varied from approximately 2.5 to 3.0 V when the external potentials varied from 10 to 60 V.

3. For the type of solid-electrolytic solution system investigated, it appears that for routine testing of chloride diffusivity, a simple correction factor of 2.5 V for the potential drop in the migration system outside the test specimen may be introduced.

References

1. O. E. Gjørv, *V. Mohan Malhotra Symp. on Concrete in the 21st Century*, Ed. P. K. Mehta. San Francisco, ACI SP-144, 1994, pp.545–574.

2. O. E. Gjørv and K. Sakai, *Concrete Under Severe Conditions: Environment and Loading, Vol. 1*, Eds K. Sakai, N. Banthis and O. E. Gjørv. E & FN Spon, London, 1995, pp.655–666.

3. Nordtest Method (1989), NT BUILD 355, ISSN 0283-7153, 1989.

4. C. Andrade and M. A. Sanjuàn, *Adv. Concr. Res.*, 1994, **23**, (6), 127–134.

5. P. F. McGrath and R. D. Hooton, *Cem. Concr. Res.*, 1996, **26**, (8), 1239–1244.

6. T. Zhang, (1997), Doctoral thesis 1997: 132 Department of Building Materials, Norwegian University of Science and Technology, Trondheim, Norway.

7. T. Zhang and O. E. Gjørv, *Cem. Concr. Res.*, 1994, **24**, (8), 1534–1548.

2

Results From Laboratory Tests and Evaluation of Literature on the Influence of Temperature on Reinforcement Corrosion

M. RAUPACH

Schießl • Raupach Consulting Engineering,
Bergdriesch 4, D-52062 Aachen, Germany

ABSTRACT

The large number of corrosion problems in reinforced concrete structures worldwide has led to the durability of concrete structures exposed to aggressive environments becoming a problem of major importance. New results from research on chloride-induced corrosion of steel in concrete show that the corrosion mechanisms are quite complex. Normally, locally separated anodically and cathodically acting areas are formed on the steel surface. As the cathodically acting steel surface areas are not visible, the corrosion mechanisms can only be investigated indirectly using new electrochemical testing methods. In this paper the influence of temperature on the corrosion rate is discussed based on the theoretical background of chloride-induced macrocell corrosion of steel in concrete.

1. Introduction

During recent years, corrosion of the reinforcement induced by chlorides has caused serious damage to concrete structures all over the world. Bridges and car parking structures have been damaged by the use of de-icing salts so seriously that many of them have had to be repaired or replaced. Offshore structures like jetties, piers, dams, docks or harbour structures are also attacked by chlorides from seawater, especially in the tidal, splash and spray zones.

This paper is concerned only with chloride-induced corrosion; the theoretical background will be discussed first and then the influence of temperature.

2. Theoretical Background
2.1. General

Chlorides can penetrate into concrete that is in contact with de-icing salts or sea water. In exceptional cases critical amounts of chlorides may also be present in the fresh concrete mix or penetrate into the concrete following fires involving PVC (polyvinyl chloride) components and structures.

2.2. Electrochemical Corrosion Process

Chloride-induced corrosion of steel in concrete is an electrochemical process comparable to the operation of a battery. The poles of the battery are different surface areas of the reinforcement acting anodically or cathodically:

- At the anode, iron ions pass into solution, leaving assembled electrons in the metal. The ions are converted into rust products in further reactions.

- At the cathode, electrons, water and oxygen are converted into hydroxyl ions. The cathodic process does not cause any deterioration of the steel but, on the contrary, can result in its protection.

- These hydroxyl ions with their negative charge move in the electrolyte through the electrical field created between anode and cathode, towards the direction of the anode. Near the anode, they react with the iron ions in solution. Depending on the moisture (i.e. in conditions where continuous full immersion does not occur) and aeration conditions, this intermediate product may continue to react, producing the final corrosion products. The individual processes are in fact much more complicated.

2.3. Simplified Electrical Circuit Model

Based on a simplified electrical circuit model [1] the relationship between driving voltage, the resistances of the corroding system and the electrical macrocell current which is proportional to the corrosion rate, can be calculated. This model consists of the driving voltage U_e, the resistances of steel, anode, cathode and electrolyte (concrete) and the electrical current I_e flowing between anode and cathode. The galvanic current can be calculated using the following basic equation:

$$I_e = \frac{U_{R,c} - U_{R,a}}{\dfrac{r_a}{A_a} + \dfrac{r_c}{A_c} + k \times \rho_{el}} \tag{1}$$

where:

I_e	=	electrical current between anode and cathode (corrosion rate)
$U_{R,c}$	=	rest potential at the cathode
$U_{R,a}$	=	rest potential at the anode
r_a	=	specific anodic polarisation resistance
A_a	=	anodically acting steel surface area
r_c	=	specific cathodic polarisation resistance
A_c	=	cathodically acting steel surface area
ρ_{el}	=	specific resistance of the electrolyte (concrete)
k	=	cell constant (geometry).

The resistance of the steel related to the transport of electrons is negligibly small compared to the other active resistances. The resistances of anode, cathode and electrolyte can again be sub-divided into different partial resistances. Besides the anodic, cathodic and electrolytic processes the geometry of the corrosion cell also influences the corrosion rate directly.

2.4. Basic Influencing Parameters

The parameters influencing the corrosion rate of macrocells in concrete structures are schematically shown in Fig. 1.

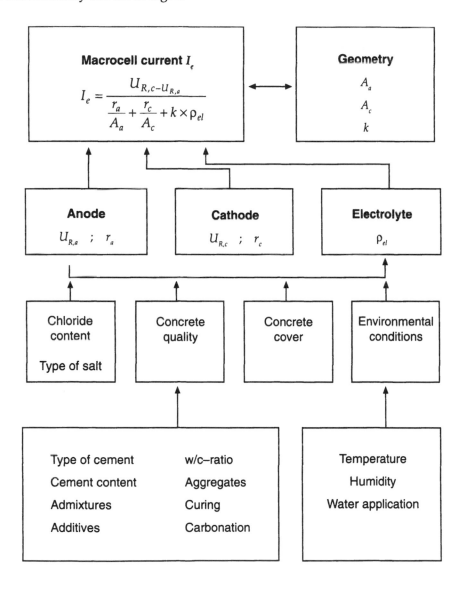

Fig. 1 *Factors influencing the corrosion rate in chloride induced macrocell corrosion.*

It is evident that a prediction of the macrocell current is quite difficult due to the large number of influencing factors. Additionally, the relationships between the parameters and the three resistances are different: for example, water saturation of the concrete leads to a very low electrolytic resistance and a high cathodic resistance.

2.5. Influence of Temperature

According to eqn (1) the corrosion rate depends on the driving force U and the resistances of anode, cathode and electrolyte. To quantify the influence of temperature on the corrosion rate, the relationships between temperature and all variables of eqn (1) (U, r_a, ρ_{el}, r_c) have to be known.

The fundamental relationship between an electrochemical process and temperature can be described by the following equation:

$$I = \frac{I_o}{e^{a(1/T - 1/T_0)}} \qquad (2)$$

where:

I	=	current at temperature T
I_0	=	current at temperature T_0
a	=	activation energy constant in K
T, T_0	=	absolute temperatures in K.

On the other hand, the influence of temperature on resistance can be described by the following equation:

$$R = R_0 \; e^{b(1/T - 1/T_0)} \qquad (3)$$

where:

R	=	resistance at temperature T
R_0	=	resistance at temperature T_0
b	=	activation energy constant in K
T, T_0	=	absolute temperatures in K.

For the activation energy constants a and b a value of 1 K corresponds to 8.314 J/ mol. In the following sections the influence of temperature on the different parameters is discussed in detail.

3. Results from Laboratory Tests
3.1. General

As most laboratory tests have been carried out at constant temperature, there are only a few results from the literature on the influence of temperature on the parameters influencing the corrosion rate. Most of these tests have been carried out to investigate the influence of temperature on the electrical resistivity of concrete.

In the following the results from laboratory tests on the influence of temperature on the corrosion rate based on macrocell current measurements are first described and then those results from further laboratory tests and on-site measurements, which have been found in the literature.

3.2. Laboratory Investigations Based on Macrocell-Measurements

3.2.1. General
The investigations described in the following sections have been carried out as part of an extensive research program at the Institute for Building Materials Research of the Technical University of Aachen, ibac, Germany on the mechanisms of, and parameters influencing, chloride induced corrosion of steel in concrete.

3.2.2. Test Set-up and Materials
The tests have been carried out in a polymer vessel containing concrete with a chloride content of 2 % relative to the cement content which was added directly to the mixing water. The cross section of the concrete was 33 cm × 25 cm and the thickness was 10 cm (see Fig. 2). The separation of the two macrocells was 15 cm and the concrete cover to the free concrete surface was 13 mm (see Fig. 2).

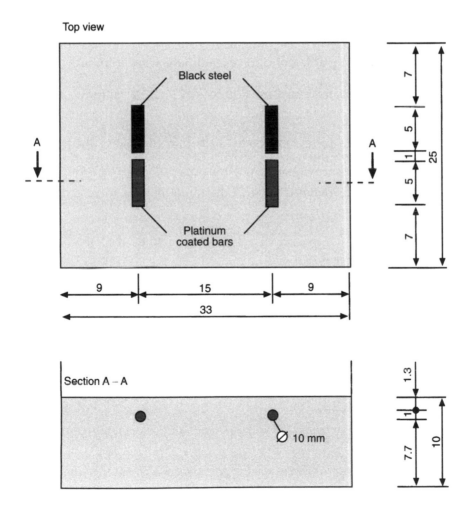

Fig. 2 Set-up of the laboratory test specimen.

The concrete composition was chosen to be 300 kg/m³ OPC 35 and water/cement-ratio = 0.6. Gravel with a maximum aggregate size of 16 mm was used. The aggregate grading was in accordance with the grading curve limits A16/B16 of DIN 1045. The concrete was cured in a fog chamber for two days.

As anodes two pieces of black steel with a diameter of 10 mm and a length of 5 cm were used (steel surface = 15 cm²) and as cathodes platinum oxide-coated titanium bars of the same size, these being protected against corrosion in this concrete by a passive layer on the surface. The platinum oxide-coated bars were used to get a high driving voltage between anodes and cathodes so as to simulate a large area of non-corroding passive reinforcement bars near the anodes — as is often the case in practice.

3.2.3. Testing Programme

The test specimen was stored in a climatic chamber at different temperatures and relative humidities (altogether seven cycles). During these different storage conditions the electrical macrocell currents between one cell consisting of anode and cathode were measured continuously using a computer controlled measurement system. Additionally, the electrolytic resistance of the concrete between the second anode and cathode were also measured continuously. The results of the measurements are plotted in Fig. 3. The seven different storage conditions were as follows (see Fig. 3):

1. The temperature was increased from 20 to 35°C; the humidity was kept at 60% R.H.

2. The temperature was increased step by step from 20 to 63°C; the humidity was kept at about 70% R.H.

3. The temperature was reduced to 20°C; the humidity was increased to 88% R.H.

4. The temperature was increased step by step from 20 to 61°C; the humidity was kept at about 88% R.H.

5. The temperature was decreased back to 20°C and then increased from 20 to 40°C; the humidity was kept at 50% R.H.

6. The temperature was increased to 60°C; the humidity was kept near 100% R.H.

7. The temperature was decreased to 20°C; the humidity was reduced step by step from 100 to 50% R.H.

3.2.4. Test Results and Discussion

The test results are shown in Fig. 3. As expected the macrocell current normally increased with temperature while the electrolytic resistance of the concrete decreased.

One exception was the result where the macrocell currents did not increase during the increase of temperature from 20 to 35°C at the beginning of the testing programme. This result can be explained by drying out of the specimen.

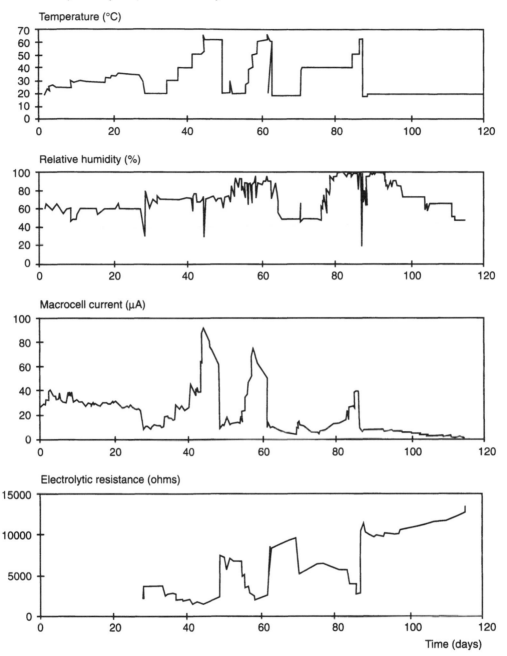

Fig. 3 *Results from laboratory tests. Macrocell currents and electrolytic resistance of concrete of test specimen exposed in a climatic chamber at various temperatures and relative humidities.*

Another interesting effect was that the electrolytic resistance of the concrete was much higher at the end of the testing programme than at the beginning although the temperature and humidity of the air were nearly the same. This effect can also be explained by drying out of the specimens or perhaps additionally by changes in the pore structure of the concrete resulting from storage under high temperatures up to 63°C.

For the results from cycles No. 2 and 4 the constants of eqns (2) and (3) were calculated.

The calculation of the coefficient b of equation (3) showed that the Arrhenius eqn was suitable to describe the relationship between the electrolytic resistance of the concrete and temperature. The results from literature and from the laboratory tests described above are shown together in Fig. 4.

The relationship between the macrocell currents and temperature did not follow the Arrhenius equation as well as that between temperature and electrolytic resistance. As an approximation, therefore, the a coefficients have been calculated as the best fit to eqn (2). The results are shown in Fig. 5.

For the cycles No. 2 and 4 the coefficients fitting best were $a = 3879$ K at 70 % R.H. and 4835 K at 88% R.H. This result indicates that the influence of temperature on the corrosion rate depends significantly on the air humidity and subsequently on the water content of the concrete.

Compared to the coefficients a for the influence of temperature on the corrosion rate the influence of temperature on the electrolytic resistance of the concrete characterised by the coefficient b is significantly less. This means that the corrosion rate depends more on temperature than on the resistance of the concrete. As mentioned before this can be explained by the fact that the potential difference between anode and cathode and the polarisation resistances also influence the corrosion rate as well as the electrolytic resistance of the concrete.

To clarify this problem the potential difference between anode and cathode was measured after the testing programme. It was found that these driving forces were not constant but increased from 155 to 230 mV when the temperature was raised from 20 to 60°C. This relationship might provide at least one reason for the result

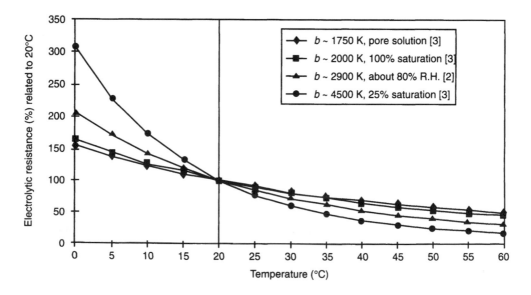

Fig. 4 Influence of temperature on resistivity of concrete.

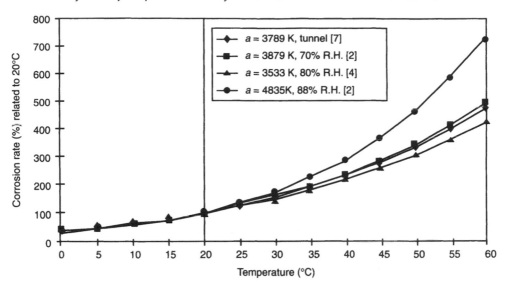

Fig. 5 *Influence of temperature on the corrosion rate of chloride induced macrocell corrosion of steel in concrete.*

that the corrosion rate depends more on temperature than on the electrolytic resistance of the concrete.

3.3. Results from Other Laboratory Tests

3.3.1. Influence of Temperature on the Electrical Resistivity of Concrete

Several results of laboratory tests on the influence of temperature on the electrical resistivity of concrete are summarised in the literature [2,3]. For normal concrete, which is not water saturated and not totally dried out, the activation energy constant b is in the range of 3000 K.

Results from tests at different degrees of saturation have been described [3]. They show, that the activation energy constant decreases significantly, when the degree of water saturation increases, thus, it was found, that the constant b was in the range of 4500 K at 20% saturation, in the range of 2500 K at 80% and about 2000 K in the range of 100% water saturation. Additionally it has been found, that the activation energy constant b of the pore solution is in the range of 1750 K [3]. The relationships between resistivity and temperature calculated from these activation energy constants are shown in Fig. 4.

This effect, that the activation energy increases with reduced degree of water saturation, has been explained as follows [3]. When the water content is reduced, the pores are less filled with water, leading to thin layers of water on the inner walls of the pores. As it can be assumed that the activation energy of the double layer is much higher than the activation energy of the pore solution, the activation energy constant consequently increases with decreasing degree of water saturation.

3.3.2. Influence of Temperature on the Corrosion Rate

The results of Bertolini and Polder [4], carried out at 80% R.H. between about 13 and 30°C lead to an increase of the corrosion rate by up to a factor of 2 corresponding to an activation energy constant a of 3533 K. The results are plotted in Fig. 5 as curves of corrosion rate against temperature.

However, Fig. 5 shows the unfavourable situation of regions with a hot climate. Thus, compared to a temperature of 20°C the corrosion rate of the reinforcement is twice as high at about 35°C. At higher temperatures the corrosion rate increases more than proportionally.

3.3.3. Influence of Temperature on the Cathodic Process

Investigations on the effect of temperature on oxygen transport through submerged concrete [5] have shown, that the cathodic current-density of oxygen reduction increased from 0.5 mAm^{-2} to 1.2 mAm^{-2} for a temperature increase from 1 to 30°C. This suggests an activation energy constant a of about 2500 K.

Therefore, to estimate the influence of temperature on the corrosion, it is generally of importance to determine which resistance is the decisive parameter, i.e. R_a, R_c or R_{el}.

4. Results from Measurements On-Site

Results taken from the Anode-Ladder-System for corrosion monitoring as described for example by Schießl and Raupach [6] show, that the constant b for the influence of temperature on resistivity is in the range between about 2900 K and 3300 K, which is in agreement with the results described in section 3.2. from laboratory tests.

Results described by Elsener *et al.* [7] from the St. Bernardino Tunnel in Switzerland showed, that an increase in temperature from –10 to +18°C led to an increasing current-density of a cathodic protection system from 5.5 to 22 mAm^{-2}. This corresponds to an activation energy constant $a = 3789$ K, which agrees well with the results from the laboratory tests described in section 3.2.

Corrosion measurements on concrete structures in Spain have shown that the influence of temperature on the corrosion rate is considerably dependent on humidity and that the corrosion rate can decrease with increasing temperature due to the effect of drying out of the concrete under the specific environmental conditions [8].

5. Conclusions

The following conclusions can be drawn from the results described in this paper:

- The influence of temperature on corrosion rate cannot generally be described by a constant activation energy, as it is dependent on several parameters, especially the degree of water saturation of the concrete. Therefore, a higher temperature does not in principle cause increasing corrosion rates because of the possible overlapping effect of drying out of the concrete.

- The influence of temperature on the resistivity of the concrete depends significantly on the water saturation of the concrete: the activation energy constants vary between about 4500 K to about 2000 K, when the degree of saturation increases from 20 to 100%.

- The influence of temperature on the corrosion rate is not only influenced by the effect of temperature on the resistivity of the concrete, but also by the resistances of anode and cathode. For concrete which is not submerged the activation energies are often in the following ranges:

  Activation energy constant a for the corrosion rate: about 4000 K;
  Activation energy constant b for resistivity: about 3000 K.

- A voltage measurement has shown, that the driving voltage between anode and cathode of an artificial active macrocell increases significantly with temperature [2]. Further research is needed to verify this effect.

- Investigations on the effect of temperature on oxygen transport through submerged concrete have shown, that the oxygen transport depends on temperature with an activation energy of about 2500 K.

- In literature no results from systematic studies on the influence of temperature on the anodic polarisation resistance and the cathodic polarisation resistance have been found.

6. Further Outlook

To be able to estimate the influence of temperature on corrosion rate, for example, to estimate the durability of concrete structures or to interpret electrochemical corrosion measurements correctly, systematic studies on the effect of temperature on the different parameters influencing the corrosion rate are still needed. As there are already several results on the effect of temperature on resistivity of the concrete, the emphasis should now be placed on studies of the effects of temperature on the anodic and cathodic reactions.

References

1. M. Raupach, Chloride-induced macrocell corrosion of steel in concrete – Theoretical background and practical consequences, *Const. Build. Mater.*, 1996, **10**, (5), 329–338.
2. P. Schießl and M. Raupach, Influence of temperature on the corrosion rate of steel in concrete containing chlorides, in *1st Int. Conf. on Reinforced Concrete Materials in Hot Climates*, Abu Dhabi, 1994. Publ. United Arab Emirates Univ., College of Engineering, P.O. Box 17555, Al Ain, UAE.
3. D. Bürchler, Der elektrische Widerstand von zementösen Werkstoffen. The Electric Resistance of Cement Based Materials. Dissertation ETH Zürich, Nr. 11 876, Zürich 1996 (in German).
4. L. Bertolini and R. Polder, Concrete Resistivity and Reinforcement Corrosion Rate as a Function of Temperature and Humidity of the Environment. TNO report -97-BT-R 0574, TNO Building and Construction Research, 1997.

5. O. Vennesland, Effect of temperature on oxygen transport through submerged concrete. *Nordic Concr. Res.*, 1991, **10**, 139–145.

6. P. Schießl and M. Raupach, Monitoring system for the corrosion risk of steel in concrete. *Concr. Int.*, 1992, **7**, S. 52–55.

7. B. Elsener, D. Flückiger, H. Woytas and H. Böhni, Methoden zur Erfassung der Korrosion von Stahl in Beton. Methods to determine the corrosion of steel in concrete. Eidgenössisches Verkehrs- und Energiewirtschaftsdepartement, Bundesamt für Straßenbau, Zürich, February 1996, No. 521 (in German).

8. C. Andrade, C. Alonso and J. Sarra, Dependence of corrosion rate values on climatic parameters in aerial outdoor concrete structures contaminated with chlorides. This volume, p. 83–91.

3

Corrosion Induced Cracking of Reinforcing Steel

U. NÜRNBERGER

Research and Testing Institute, Baden-Württemberg, Stuttgart, Germany

ABSTRACT

In concrete constructions stress corrosion cracking and corrosion fatigue of reinforcement may occur. Stress corrosion cracking of low carbon reinforcing steel is the result of the influence of nitrate. In carbonated concrete the resistance to SCC is lower than in alkaline concrete. The sensitivity of the steel increases with decreasing carbon content.

In dynamically stressed cracked concrete constructions under sea water exposure (e.g. offshore constructions) corrosion fatigue may take place. Permanently immersed parts are more resistant than those occasionally sprayed. Epoxy coating cannot protect the steel sufficiently because the coating may be damaged as a consequence of fretting.

1. Introduction

In reinforced concrete rebars corrode uniformly in carbonated concrete or by pitting in chloride-enriched concrete irrespective of the stress condition. The reasons are low cover and quality of concrete. As a result of these corrosion phenomena, spalling of concrete and reduction of cross section may occur. Nevertheless, load and corrosion induced cracks are more detrimental, since these may lead to stress corrosion cracking and corrosion fatigue.

2. Stress Corrosion Cracking (SCC)

In the 1960s in Germany some 25-year old reinforced concrete ceilings of pigsties collapsed. The reasons for this were intercrystalline SCC of low carbon reinforcing steel (C ≤ 0.05%) (Figs 1 and 2) as a consequence of nitrate influence. The mortar around the steel contained up to 0.15 wt% nitrate; the concrete cover was always carbonated.

Fig. 1 SCC of low carbon reinforcing steel.

Fig. 2 *Intergranular SCC of low carbon reinforcing steel.*

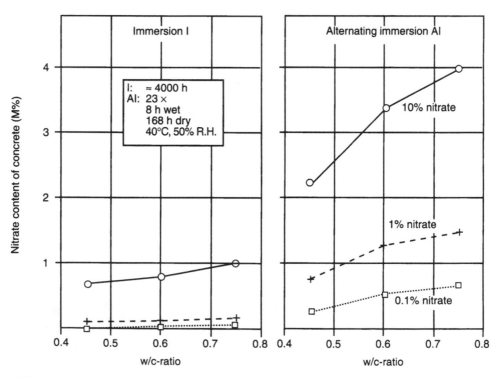

Fig. 3 *Nitrate content of concrete at 1 cm depth after treatment with solutions of different concentrations.*

Nitrates in concrete result from storage of fertilisers and the presence of nitric oxides in the surrounding area of the construction. In cattle stables calcium nitrate $(Ca(NO_3)_2)$ is produced by bacterial processes of components of liquid manure. Nitrates in the dissolved condition penetrate into the pore system of the concrete. During alternating wetting (absorption) more nitrates may penetrate into the interior of concrete than during immersion (diffusion) (Fig. 3).

Results of stress corrosion tests on reinforced concrete rebars in nitrate solution were published previously by the author [1,2], and, further, more practical investigations were conducted in cattle rooms [3]. It was found that the sensitivity to SCC of carbon steel in nitrate solution is caused by the special tendency to passivation in this solution. In concrete, passivity due to dissolved nitrate ions is also influenced by the pH-value of the pore solution (Fig. 4): in alkaline media the resistance to SCC is much higher than in the neutral range. Recorded failures have shown that nitrate induced stress corrosion cracking is a special problem with carbonated concrete.

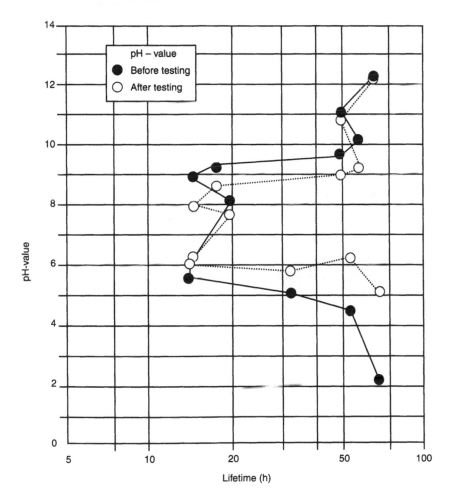

Fig. 4 *Lifetime of reinforcing bar in SCC test with respect to pH value of boiling 10% calcium nitrate (stress = 0.7 R$_p$ 0.2).*

Intercrystalline SCC is a selective type of corrosion with accelerated metal dissolution at the crack tip (grain boundary). In the case of low carbon steel the cathodic effectiveness of the grain faces is increased by the cementite layers (Fig. 5). This explains the high sensitivity of low carbon steels (see below).

The influence of environmental parameters such as concentration and temperature were tested earlier [2]. It was found that with decreasing concentration C, temperature T of the nitrate solution and mechanical stress σ_0 the lifetime, L, until fracture takes place, increases. The following mathematical relation exists between these parameters (c_0 is the resistance of the material against SCC):

$$L \cdot \sigma_0^n \cdot \sqrt{C} \cdot e^{\frac{-Q(C,\sigma,T)}{RT}} = c_0$$

It follows that SCC may occur after about 20 years, if the material is sensitive, the stress is about 60% of the yield strength and the NO_3^{2-} concentration of pore solution is about 1 wt%.

If the carbon content is lower than 0.2% the sensitivity of the reinforcing steel increases with decreasing carbon content. If this is less than or or equal to 0.05% or the structure has decarburised zones, the sensitivity to nitrate-induced SCC reaches a maximum (Fig. 6). Furthermore, high contents of nitrogen and phosphorus are unfavourable. This explains why low carbon Thomas steel of older production was particularly prone to failure.

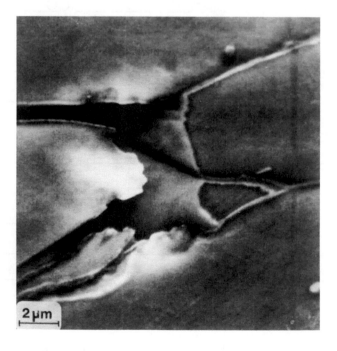

Fig. 5 *Cementite layers on grain boundaries of low carbon steel.*

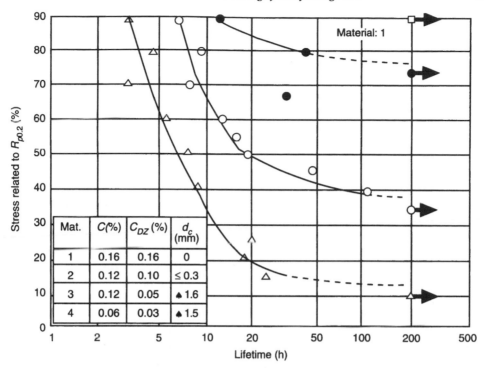

C_{DZ} = carbon content of decarburised zone.
d_c = depth of decarburised zone.

Fig. 6 *SCC-behaviour of different hot rolled ribbed reinforcing bars in boiling 30% calcium nitrate solution.*

No considerable external stress is necessary to cause stress corrosion cracking. Internal tensile stresses resulting from production (twisting) or processing (bending) are sufficient to produce SCC in sensitive steels.

3. Corrosion Fatigue

Constructions may also be subjected to dynamic influences, e.g. wind, sea waves or road traffic. Dynamically stressed reinforced concrete constructions are cracked and corrosive media (deicing salts, sea water) and oxygen have relatively free access to the reinforcement. Therefore constructions have to be designed with corrosion fatigue in mind.

In bridges as well as in offshore constructions stresses are of low frequency but lead to high stress amplitudes. For example, in offshore constructions, the highest waves with the most effective stresses occur with a low frequency of 0.1 s^{-1} leading to especially unfavourable conditions with regard to corrosion fatigue. Therefore corrosion fatigue tests on cracked reinforced concrete beams have been conducted with realistic sea water exposure and low frequency. The experimental details and further results have been reported [4].

In chloride-containing concrete corrosion fatigue (Fig. 7) takes place in an active state. Preferred initial points of precracks are wide corrosion pits (Fig. 8) which result from the attacking chloride ions. Here the notch effect is increased and these areas are sensitive to crack initiation at slip bands. Chloride levels are increased and enriched by electrolytic migration and iron chlorides hydrolyse to provide an acid electrolyte. Because during crack initiation and propagation corrosion intensity is time dependent, low frequencies will favour corrosion fatigue.

Cracked concrete beams reinforced with unprotected and epoxy coated ribbed bars BSt 500 S (German standard) were tested in air and sea water conditions in fatigue tests with pulsating bending stresses at a frequency of 0.5 s^{-1}. The beams were either sprayed daily (5 min per day) with artificial sea water or were permanently immersed. Some test results are summarised in Fig. 9. The reinforced beams treated with sea water showed significantly more unfavourable behaviour than those tested in air (the fatigue strength in air is about 250 Nmm^{-2}). Permanently immersed beams were more resistant than those sprayed periodically. This difference in behaviour of the reinforcing steels in immersed and sprayed conditions is the result of the differing rates of oxygen transport through the concrete crack to the steel, since water in the crack hinders oxygen diffusion.

Beams with epoxy coated reinforcement are no better than beams reinforced with unprotected steels. In the crack area the coating is destroyed as a result of fretting. This leads to crevice corrosion, rusting and acidification in the crevices as a result of hydrolysis of corrosion products. In other fatigue tests on concreted bars [5,6] it was stated that corrosion protected bars behave better than unprotected.

Fig. 7 *Reinforcing steel after corrosion fatigue test.*

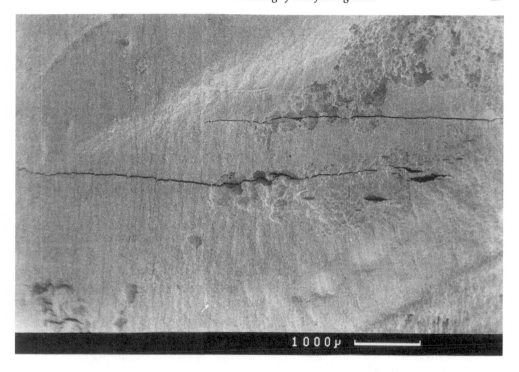

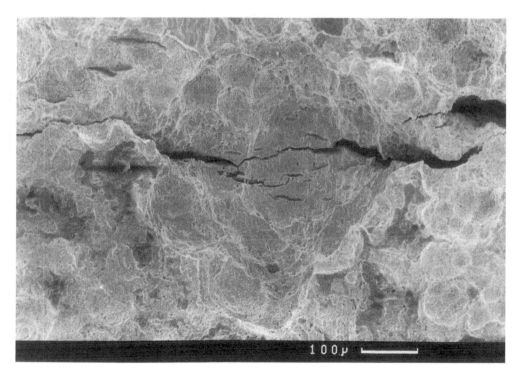

Fig. 8 *Pitting induced corrosion-fatigue in chloride-containing concrete.*

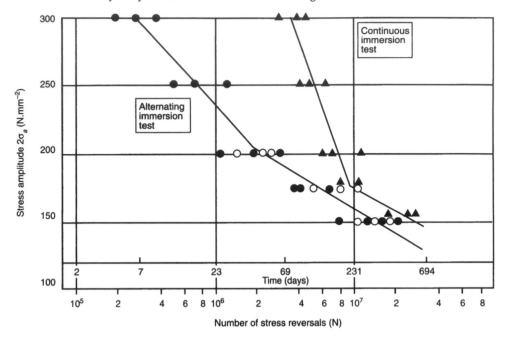

Fig. 9 *Fatigue tests on cracked reinforced concrete beams in sea water (frequency 0.5 s⁻¹).*
● ▲ Steel not protected; o Steel epoxy coated.

Applying this knowledge to dynamically loaded structures in sea water, it becomes clear that static systems with the greatest alternating load in the tidal zone or directly above are most endangered.

References

1. U. Nürnberger, *Arch. Eisenhüttenwes.*, 1973, **44**, 775–784.
2. U. Nürnberger, *Werkst. Korros.*, 1977, **28**, 312–321.
3. G. Rehm, U. Nürnberger and N. V. Waubke, *Beton. Fertigteil-Technik*, 1973, 642–650.
4. U. Nürnberger and W. Beul, Research Report FMPA BW Stuttgart/Germany, 1997, Vol. 8,
5. T. Hodgkiess, P. D. Arthur and J. C. Earl, *Mater. Perform.*, 1984, **23**, 27–31.
6. M. Howkins, *Mater. Constr.*, 1984, **17**, 69–74.

Part 2

Reference Electrodes, Supplementary Protection Measures

4

Junction Potentials at a Concrete/Electrolyte Interface

H. ARUP and O. KLINGHOFFER*

Hans Arup Consult, DK-5600 Faaborg, Denmark
*FORCE Institute, DK-2605 Brøndby, Denmark

ABSTRACT

Examples are given of junction potentials in concrete. Diffusion, streaming and membrane potentials contribute to the variability and time dependency of surface potentials, but are difficult to calculate. A viscous contact electrolyte significantly changes the junction potential. The concept of a 'true potential' — so far a *fata morgana* — is introduced and discussed.

1. Introduction

The first publication on the monitoring of steel corrosion in concrete by potential mapping is now 40 years old, but we still have difficulties in defining and measuring 'true' potentials in concrete because of the variability of the junction potential across the interface between concrete and the electrolyte of the reference electrode. The interpretation of surface potential gradients as being caused by IR drops of macrocell currents is only meaningful if it can be assumed that junction potentials are absent or constant in value and independent of changing surface conditions.

In an earlier publication [1] one of the present authors tentatively identified three possible electrochemical phenomena, which might contribute to the junction potential at a concrete/electrolyte interface:

- Diffusion potentials;

- Membrane potentials; and

- Streaming potentials.

It was also suggested, that additional and more fundamental work was needed, and such work has recently been forthcoming [2,3].

In preparation for the anticipated research it was decided to develop a new embeddable reference electrode, the MnO_2-electrode, which is believed to have negligible junction potentials when embedded in normal concrete [4] The embedded electrode has excellent stability and can be used as a reference against which surface-held electrode arrangements are measured [5].

It is the purpose of this chapter to discuss recent results obtained with this technique in the authors' own laboratories and elsewhere.

First, however, it might be useful to discuss the concept of a 'true potential' in concrete.

1.1. What is a 'True Potential'?

The recent interest in junction potentials is — perhaps unconsciously — driven by a wish to know the 'true potential' of the embedded metal or the 'true' IR drop generated by currents flowing through concrete and across concentration gradients.

A 'true potential' may be defined as the potential which is valid in a potential/pH diagram and can be used to give precise predictions of electrochemical reactions, such as the onset of hydrogen evolution on a steel surface as a function of potential and porewater chemistry.

The definition and measurement of 'true potential' is also a prerequisite for the use of ion-selective electrodes and gas-electrodes in concrete. Examples of such electrodes are Ag/AgCl-electrodes for chloride, gold or platinised Ti as redox-electrodes and metal/metal–oxide electrodes for pH. All these electrodes have to be measured against a reference electrode and the reading converted to a 'true potential'.

In water saturated, non-carbonated concrete apparently reliable measurements can, with a minimum of precautions, be made against conventional embedded or surface-held electrodes and the reading converted to the hydrogen scale as used in a potential/pH diagram. Another recommended method is to take measurements at the bottom of a hole drilled through the carbonated or dried out surface layer. This method also has the advantage that measurements can be made close to a selected area of the embedded steel or close to an embedded electrode for the purpose of control.

Increasing difficulties are met if measurements are attempted on surfaces which have strong compositional gradients (humidity, pH, chloride) close to the surface.

The greatest problems, both in the selection of measuring techniques and in the definition of potentials, occur if the object to be measured, be it embedded metal or an ion-selective electrode, is surrounded by semidry concrete. In semidry concrete the embedded electrode is in contact with liquid in the finest pores only. The composition of the electrolyte in these fine pores is virtually unknown and it has electrochemical properties different from that of the same electrolyte in bulk. Membrane potentials probably exist across the interface between the electrode and the electrolyte in the fine pores, but they are difficult or impossible to measure.

In the following section examples will be given of measurements, which illustrate various junction potential effects.

2. Experimental
2.1. Diffusion Potentials

Some users of MnO_2-electrodes have from time to time developed their own test programmes for reference electrodes that were originally meant for use with Ag/AgCl-electrodes. One of these tests used sea water as the connecting electrolyte and this produced unexpected results which have been modelled in the following laboratory experiment.

Small 15 mm dia. rods of cement paste, similar to the plugs used in the MnO_2-electrode, were cast and cured for a few days in a 0.5M NaOH. They were then mounted in heat shrinkable tubing, so a small cavity was formed over the top (Fig. 1). Saturated KCl was poured in the cavity and a Saturated Calomel Electrode (SCE-2) dipped into the solution. The lower, protruding end of the stick was dipped into a test solution, which was either 0.6M NaCl or synthetic sea water (SSW). Another SCE-electrode (SCE-1) was dipped in the same solution. The potential difference SCE-2 – SCE-1 was measured after 1 s and with increasing intervals up to 24 h. The tests were repeated with paste rods, which were superficially dried in laboratory air for 24 h before being mounted (named 'wet' and 'dry' respectively). All tests were done in triplicate.

The results, shown in Table 1, showed that the junction potential of SSW is markedly different from that of a plain NaCl-solution, probably because of the buffer capacity of SSW, which is expected to give a sharper pH-gradient across the cement electrolyte interface. Another possibility is that a precipitate of insoluble salt in the pores of the cement paste could change the ability of the interface to act as a semipermeable membrane. The 'drying' of the paste altered the initial measurements, but not the final reading.

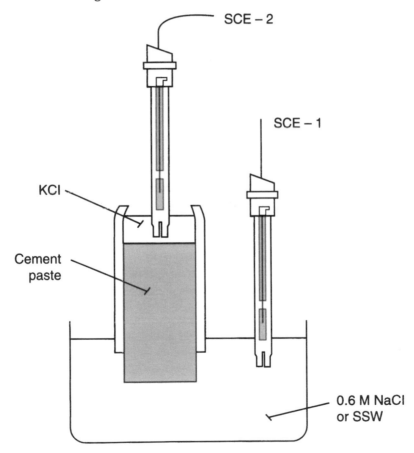

Fig. 1 *Experimental arrangement used to measure diffusion potentials.*

Table 1. Potential difference (SCE-2 – SCE-l) obtained with the arrangement shown in Fig. 2. Results are given in mV and are the averages of three tests. See also text

Solution		1 s	10 s	1 h	2 h	24 h
NaCl	'dry'	19.2	8.3	9.0	21.9	22.8
NaCl	'wet'	7.9	10.6	16.2	19.6	19.9
SSW	'dry'	35.5	11.3	9.7	38.3	44.6
SSW	'wet'	13.2	28.3	41.9	52.1	53.61

An additional experiment, not reproduced here, was made with a paste made up with SSW as mixing water. With this paste NaCl and SSW gave similar results.

2.2. Streaming Potentials

For the next experiments four blocks of OPC mortar, $30 \times 15 \times 8$ cm, each with two embedded MnO_2 electrodes, were available. Two of the blocks were diamond-cut to a thickness of 4 cm. The two thick blocks had been stored under dry indoor conditions for 20 months, the two 4 cm blocks had been outdoors under shelter, wrapped in plastic foil, for the same period, and were then transferred to the laboratory.

Throughout the observation period the difference in potential for each pair of electrodes was from 1 to 5 mV, individual pairs being extremely stable (± 0.1 mV). One of the electrode cables in a thick block was apparently damaged after one year, and the potential difference in this block rose slowly from 3 to 7 mV. Moistening of the mortar around the damaged cable resulted in a completely false reading.

The a.c. resistance between the embedded electrode pairs had risen by a factor of 3–15 during drying. This had not affected the potential difference. The large blocks were driest, having been freely exposed all the time.

One of the large blocks (with intact cables) was placed on end in a shallow, 1 cm deep pond with saturated $Ca(OH)_2$, connected via a KCl-salt bridge to a SCE. From the moment of immersion the potentials of the two embedded MnO_2-electrodes (Ref. 3, Ref 4) were recorded against the SCE (see Fig. 2). A very substantial junction potential was noted, with a peak value after a few seconds of about +300 mV relative to the steady state value, reached after approximately 1 h. The small plateau after 2 minutes cannot be explained; the dip after 21 minutes was caused by inadvertent shorting of one electrode to the SCE. The difference in potential of the two embedded electrodes remained constant.

The next day the two 4 cm blocks were placed in the pond, resting on the long, narrow side. Seconds later the thick block, now with one wet end, was placed on two moistened pads raised above the pond, so electrolytic contact to the SCE was made from the wet end and the dry end simultaneously. The 'wet' contact dominated — as expected — and the junction of the dry end was suppressed. But the relative potentials of the two embedded electrodes (Refs 3, 4) fluctuated — including a sign

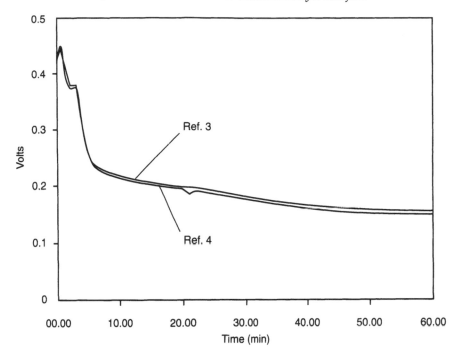

Fig. 2 *Potential of two embedded MnO₂-electrodes in dry mortar measured against SCE in Ca(OH)₂ solution as a function of time after wetting.*

reversal. This could be interpreted as resulting from a small current, probably less than 1 μA, flowing through the concrete from the 'dry' to the 'wet' junction. The resistance between the electrodes in the block was approximately 30 kΩ. The four embedded electrodes in the two 4-cm blocks showed very parallel junction potential peaks (Refs 5–8), but much smaller, in the region of 70–80 mV (Fig. 3).

The next experiment to be described here was done a year earlier on one of the large blocks, before the defect cable was in evidence. It illustrates the peculiar effect of using a viscous electrolyte. Figures 4 and 5 show first the now familiar, quickly vanishing, junction potential which developed when the dry surface was contacted with an SCE through a sponge wetted with tapwater. When the tapwater was made into 'wallpaper glue' with carboxy methyl cellulose (CMC) and a fresh contact made on a still dry area of the block, the junction potential developed much more slowly as seen in Fig. 5.

The result surprised us, because we had, probably very naively, thought that the highly viscous electrolyte would flow much more slowly through the capillaries — if at all, and that could be expected to eliminate the streaming potential. The authors admit being easily frightened in face of complicated electrochemical calculations, and knowing that researchers with more recent education have found it difficult to handle streaming and diffusion potentials in cementitious materials, we shall not attempt to explain this experiment (which has been repeated with other electrolytes). It is interesting, however, to speculate that the viscous electrolyte is in contact only with the finest pores, and that the readings reflect the properties of a material behaving like an ion-selective membrane, developing the so-called Donnan potentials.

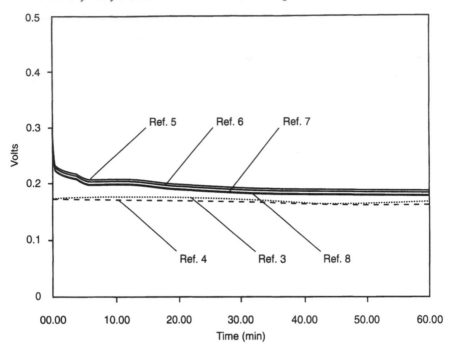

Fig. 3 *Potential of the same two electrodes (Fig. 2) 24 h later together with potentials of 4 additional electrodes (Refs 5–8) in two freshly immersed mortar blocks. See text for further details.*

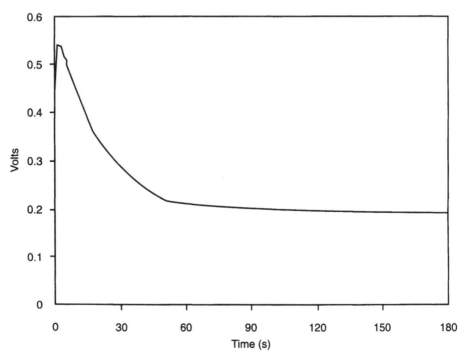

Fig. 4 *Potential of an embedded MnO$_2$-electrode in dry mortar against a surface-held SCE, using a sponge with tap water as the contacting electrolyte.*

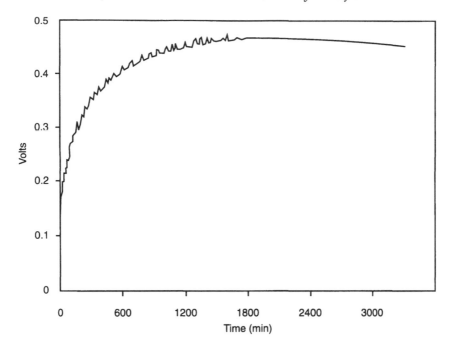

Fig. 5 *A repetition of the experiment in Fig. 4, except that the tap water is made into a slowly flowing paste with CMC.*

It is hoped that others, with a better theoretical background, will be tempted to explore this technique further. It would be natural to see if the effect is different in carbonated materials and to see how agar-stiffened gels behave compared to the viscous CMC paste. Concerning streaming potentials, it might be added, that an attempt by one of the authors to measure the potential shift across a cement paste membrane exposed to a 10 bar/cm pressure gradient was unsuccessful.

2.3. A Good Solution

Finally, we have made additional measurements on a 10 year old concrete block with several embedded MnO_2-electrodes. Early measurements on the same block were described previously [4], and it will be remembered, that there was a difference of 90–200 mV between potentials measured at the surface and potentials measured at the bottom of a hole drilled through the weathered surface. The new measurements more or less confirmed earlier observations, but in addition it was noted that a saturated solution of KNO_3 was a good contact electrolyte, which minimised the settling time of the measurement. A chloride-free electrolyte is always attractive to use, especially in drilled holes which come close to embedded steel. It reminded us of a trick developed 15 years ago by a technician charged with taking potential mappings on thousands of balconies, some of which had dirty, water repellent surfaces. The trick was to mark the location of the measuring points with a small blow from a pyramid-pointed hammer (which produced small cracks through the surface) followed by pouring a few drops of 20% HNO_3 in each mark. Seconds later stable measurements were produced with the usual water-wetted sponge and a reference electrode.

3. Discussion

There is still a lot to be learnt about junction potentials in concrete. Our experimental results are scattered and unsystematic and do not form a coherent picture. Attempts to support experimental results with electrochemical calculations have been largely unsuccessful, because mathematical solutions are only possible with simplified systems (in the case of diffusion potentials) or because the value of important parameters and constants are unknown or undefinable (streaming and membrane potentials).

There is definitely a communication gap between the cement chemists, who discuss electrochemistry in cement materials without embedded metals and use greek letters for their potentials, and the corrosion people, who expresss potential with latin letters. They should sit together and try to explain — in a common language — some of the findings mentioned above.

The concept of a 'true potential' in concrete also needs to be discussed in wider circles. The following hypothetical case illustrates the points to be discussed.

In a thick block of concrete we have a deeply embedded set of stable reference electrodes, e.g. one Ag/AgCl-electrode and one MnO_2-electrode. Another identical set of reference electrodes is placed closer to the surface together with a piece of steel and a (hypothetical) pH-electrode. The concrete becomes carbonated to such a depth that all the electrodes close to the surface are similarly affected, while the deeply embedded electrodes are unaffected. We now have a pH-gradient between the surface and the interior. Try now answering the following questions:

- Will there be a change in the relative potentials of the four reference electrodes?

- Which electrode should be used in connection with the hypothetical pH-electrode in order to monitor the progress of carbonation?

- Considering that carbonation is accompanied by drying out of the carbonated layer, how will this affect the answers given above?

- Cathodic protection is applied to the steel in the carbonated zone. Which electrode will best inform about the protection potential, seen in relation to (a) current protection criteria, and (b) risk of hydrogen damage, evaluated by reference to the Pourbaix diagram? (The effect of IR drops should not be considered.)

- Considering that CP will (1) increase the pH at the steel surface, forming a new (and steep) pH gradient in the near-surface layer, and (2) move water through the capillaries to the surface, how will these changes influence your answer to the last question?

Luckily we seem to be able to handle a lot of practical problems using CP, corrosion monitoring and potential mapping without knowing these answers. So far, so good. But knowing more about junction potentials could improve these techniques and improve the international exchange of information.

References

1. H. Arup, Electrochemical monitoring of the corrosion state of steel in concrete, in *Proc. 1st Int. Conf. on Deterioration & Repair of Reinforced Concrete in the Arabian Gulf*, Oct. 1985.

2. R. Myrdal, Phenomena that disturb the measurement of potentials in concrete, in *CORROSION '96*, paper No 339, NACE, Houston, Tx, 1996.

3. R. Myrdal, Potential gradients in concrete caused by charge separations in a complex electrolyte, in *CORROSION '97*, paper No 278, NACE, Houston Tx, 1997.

4. H. Arup and B. Sørensen, A new embeddable reference electrode for use in concrete, in *CORROSION '92*, paper 208, NACE, Houston Tx, 1992.

5. H. Arup, O. Klinghoffer and J. Mietz, Manganese dioxide reference electrode for use in concrete. This volume, p. 40–53.

5

Manganese Dioxide Reference Electrodes for Use in Concrete

H. ARUP, O. KLINGHOFFER* and J. MIETZ†

Hans Arup Consult, Denmark
*FORCE Institute, Denmark
†Bundesanstalt für Materialforschung und-prüfung (BAM), Germany

ABSTRACT

The embeddable MnO_2-electrode for use in concrete has been tested and used in the field since 1986. The present paper supplements earlier publications and presents new data on short and long term stability, polarisation behaviour, temperature response and performance behaviour in concrete with cathodically protected steel subjected to climatic cycling and field experience. Results compare favourably with published data for other embeddable electrodes and show that the MnO_2-electrode has found wide acceptance, mainly because of its very good long term stability.

1. Introduction

Embeddable reference electrodes for use in concrete are needed in connection with long-term monitoring and potentiostatically controlled cathodic protection of reinforcement in concrete. Their use in laboratory work and field exposure tests is advisable for the purpose of ensuring a valid exchange of data between laboratory and field work. As pointed out previously [1–4], they should be stable, invariant to chemical changes of the concrete, tolerant to climatic variations and have the ability to pass small currents with a minimum of polarisation and hysteresis effects.

The embeddable MnO_2-electrode for use in concrete was developed during a BRITE-programme in 1986. It is, together with the Ag/AgCl-electrode, the only embeddable true reference electrode to find widespread use in the construction industry.

A description of the electrode and its performance was published first in Scandinavia [1] and later in the USA [2]. The purpose of the present paper is to present results of additional laboratory tests and to document its long term performance. For the convenience of the reader a brief description of the electrode is given below.

The MnO_2-electrode is a half-cell consisting of a compacted mass of MnO_2 inside a metal thimble which, since 1992, has been made of stainless steel (Fig. 1). This electrode§ has been used in all the experiments described in this paper. The half cell potential is a complex function of the reduction state of the MnO_2 [5] — an earlier

---

§ Embeddable reference electrode ERE 20 from FORCE Institute.

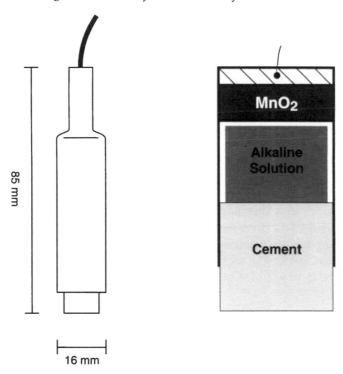

Fig. 1 *Dimensions and construction of MnO$_2$-electrode used in this work.*

statement [2] about the potential being determined by an MnO$_2$/Mn$_2$O$_3$-equilibrium was an oversimplification. With any given composition of the MnO$_2$ the potential is a linear, 59 mV/decade function of the hydrogen ion activity from pH 3.5 to over 14. The inner electrolyte is a buffered solution with a pH of approximately 13.5 corresponding to that of normal porewater. It is therefore in chemical balance with the surrounding concrete and differs from the Ag/AgCl-electrode by being chloride-free.

The electrolytic contact to the concrete is through a diffusion barrier made of fibre reinforced cement paste protruding from the insulation sheath which, in turn, protects the metal thimble and the cable connection. This ensures a very good bond to the concrete and also means that the all-important interface between the electrode and the concrete is contained within the electrode. The junction potential across this interface is minimal because the pH is nearly the same in the barrier plug and in the cell interior. Also, when the electrode is embedded in concrete, no significant junction potential develops at the plug/concrete interface. Unfortunately this construction of the MnO$_2$ reference electrode including a cementitious plug is not always well understood. If a user wishes to compare the potential of an 'as-delivered' MnO$_2$-electrode with that of a Ag/AgCl-electrode (or any other type of embeddable electrode), it will be found that the observed potential difference varies with the pH of the contact solution by as much as 30 mV/pH-unit [2,3,6]. A fairer comparison should be made with both electrodes embedded in concrete or with the comparison electrode mounted in a precast mortar block [7].

2. Experimental

In a series of laboratory tests, the new type of MnO_2-electrodes (with stainless steel thimble) has been tested for uniformity, long-term stability, temperature response, polarisation characteristics and behaviour in specimens with impressed current cathodic protection with simultaneous exposure to climatic cycles. Some of these tests were done at the FORCE Institute, others at BAM under contract with FORCE. Results of these tests were reported in 1997 [8].

2.1. Uniformity of MnO_2-Electrodes

Before delivery, the potential of every electrode is measured against a SCE (Saturated Calomel Electrode) in a thermostated solution of saturated $Ca(OH)_2$, which is maintained by daily additions of fresh $Ca(OH)_2$. The SCE is connected via a salt bridge containing KNO_3, because prolonged contact with the alkaline solutions destroys the glass frit of the SCE. Figure 2 shows the distribution of measured potentials in a batch of 172 MnO_2-electrodes.

2.2. Short-term Stability of Embedded Electrodes

The potential difference between two MnO_2-electrodes, which had been embedded in the same cement mortar block for over a year in an indoor climate, was measured with a $6^1/_2$ digit system voltmeter with 0.1 µV resolution and >10 GΩ input resistance.

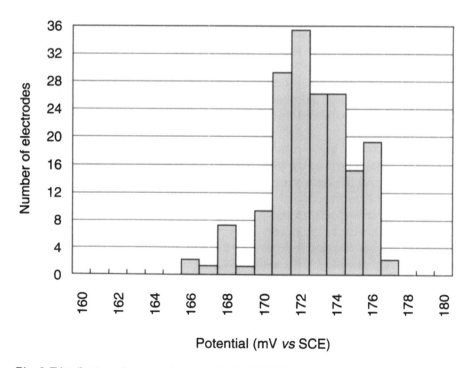

Fig. 2 *Distribution of measured potentials of 172 MnO_2-electrodes in a saturated solution of $Ca(OH)_2$.*

Readings were recorded every minute for about 3.5 h as shown in the curve in Fig. 3. There was a steady trend of approximately 6 μV h^{-1} and a ripple of about ± 2 μV.

2.3. Long-term Stability of Embedded Electrodes

Ten blocks of mortar, prepared with Ordinary Portland Cement (OPC) and quartz sand, water/cement (w/c) ratio 0.50 and with dimensions 15 ⊠ 15 × 15 cm were cast in July 1992. Each block contained a rod of 8 mm dia. smooth mild steel and 4 or 5 MnO$_2$-electrodes, 44 electrodes in total. Blocks 1–5 were stored in chambers with different relative humidities, from 32 to 100% R.H. Block 6 was exposed in air with 1.5% CO$_2$ and 60% R.H. Blocks 7–10 contained additional chloride (from 0.25–1.5% Cl of cement weight), added as NaCl to the mixing water, and these were stored at 100% R.H. All blocks were stored at room temperature.

The potential of a randomly chosen 'master-electrode' in each block was measured at increasing intervals against a SCE in contact with the surface via a sponge wetted with tap water. The potentials of the other 3 or 4 electrodes were measured relative to the master electrode. In all blocks except No. 9, the potential difference between embedded electrodes did not exceed 12 mV. In block 9, which contained 0.75% Cl, there were some unexplained aberrations, but these could be contained within a scatterband of 30 mV.

As expected from other investigations [2,6], the measurements with the external SCE were marred by the variable contributions of the junction potential. The apparent potential of the master electrodes varied from +150 to +230 mV with occasional higher

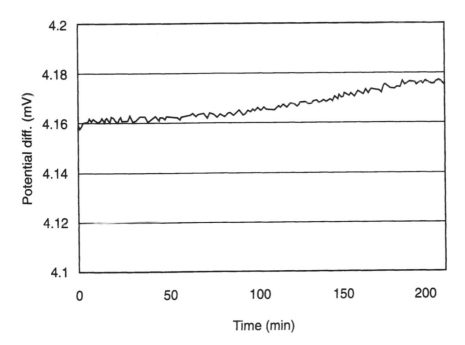

Fig. 3 *Difference in potential of two MnO$_2$-electrodes in the same block of mortar as a function of time.*

readings. This will of course make it difficult to register any trends in the potential of the embedded electrodes, and no such trends have in fact been detected during the first 4 years, except for the effect of carbonation in block 6. Fig. 4(a–c) illustrate the results from the carbonated block 6 and typical sets of curves from two of the other blocks.

2.4. Polarisation Behaviour

Polarisation is the offset in potential caused by a current flow to or from the electrode. Normal use of the electrode only gives rise to very small currents, which are in the order of 0.1 µA if the voltmeter has an input resistance of 10 MΩ. Inadvertent short-circuiting of the electrode to the steel, e.g. through an instrument used in the amp-range, results in currents only limited by the resistance in the electrode and in the concrete and possibly as high as 1 mA. Part of the polarisation is caused by the IR drop within the electrode and will disappear when the current is interrupted, but permanent changes of the potential may result if the total charge passed is large enough.

The polarisation behaviour of the MnO_2-electrode was investigated by placing two electrodes in a saturated solution of $Ca(OH)_2$ and measuring their potential relative to a SCE. They were connected in series with a 9V battery and a 1 MΩ resistor, thus passing a current between the two electrodes of approximately 9 µA. The current was maintained for 80 h. The total charge passed is 0.71 mAh or 2.6 coulomb.

Figure 5 shows the potentials of the two electrodes as a function of time during the experiment. The IR drop in each electrode is 8 mV, corresponding to an internal resistance in the electrode (mainly in the cement plug) of 900 Ω. This figure is typical for a newly made electrode; it will rise as the cement plug continues to hydrate and will attain a value of about 2 kΩ in older electrodes.

The persistent part of the polarisation after switching off the current is approximately 1.5 mV for the cathodic curve, which corresponds to loading through an instrument. If an electrode is permanently connected through a 10 MΩ instrument to cathodically protected steel at a potential of –1 V, the charge passed in 1 year will be 0.88 mAh, which may be expected to result in a permanent change of– 2 mV in the potential of the electrode. In the worst case a similar charge will have passed if the electrode is short-circuited to the steel during three quarters of an hour.

2.5. Temperature Response

There are two aspects regarding the temperature response of a reference electrode. One is the recovery of an electrode after temperature excursions, the other is the temperature coefficient of the electrode, i.e. the variation of potential with temperature. The temperature coefficient is difficult to define and to apply. The potential of the electrode under test has to be measured against a laboratory reference, e.g. SCE, and one is often left in doubt whether the laboratory reference should be kept at constant temperature or follow the temperature of the electrode under test. In the latter case, should the readings be adjusted using the literature values for the temperature coefficient of the laboratory reference?

(a)

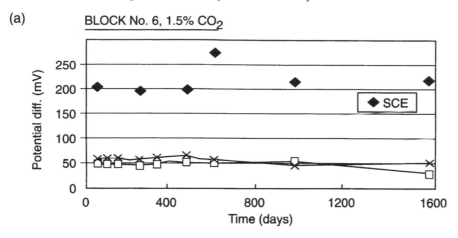

(b)

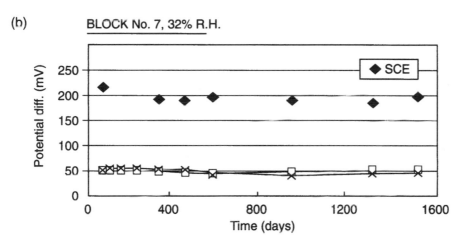

(c)

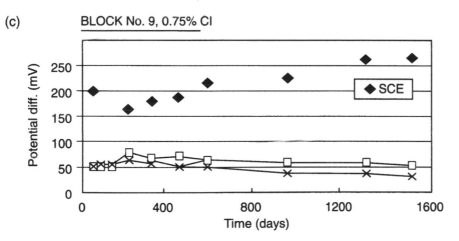

Fig. 4 (a–c) Long term relative stability of MnO$_2$-electrodes in mortar blocks at three different storage conditions; (a) carbonating atmosphere; (b) 32% R.H.; and (c) with mortar containing 0.75% chloride. See text for further explanation.

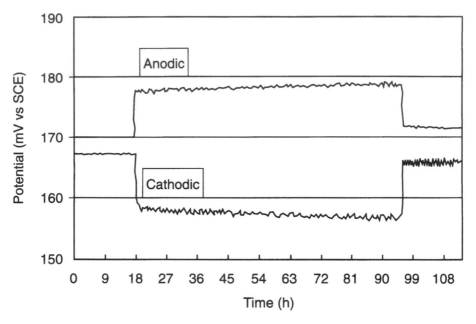

Fig. 5 *Potential changes of two MnO$_2$-electrodes while a current of approx. 9 µA was passed between them.*

The tests described below serve primarily to document the ability of the MnO$_2$-electrode to recover after thermal cycling.

In one test, ten MnO$_2$-electrodes were placed in a thermostated solution of saturated Ca(OH)$_2$ and measured against a SCE, which was immersed in the solution for 1 min before each series of ten readings. The time was probably too short for the SCE to reach a steady potential. As shown in Table 1, the ten electrodes respond very uniformly to a four day heating cycle up to a temperature of 70°C and they revert to very nearly the same potential as before heating, especially if correction is made for the small difference in temperature before and after, using the 'temperature coefficient' of –0.77 mV/°C found in this particular test.

In another temperature test, performed at BAM, three MnO$_2$-electrodes and three Ag/AgCl-electrodes (type WE50 from Silvion), the latter being delivered precast in cement mortar by the supplier, were placed in a thermostated bath of stirred, saturated Ca(OH)$_2$. Potentials were measured against a calibrated saturated Ag/AgCl-electrode, which was placed in the bath for one minute before each measurement. From the initial temperature of +5°C the temperature was raised to +10°C and then in 10°C steps to +80°C. At each step the temperature was kept constant for 4 days before the potentials were measured. Up to about 40°C the potential of the MnO$_2$-electrodes was nearly constant. At higher temperature a decreasing tendency could be observed. The potential of the Ag/AgCl electrodes showed much more scatter. After completion of the test all three Ag/AgCl cells showed cracks in the mortar surrounding the cell itself, and this might have affected the measurements (Fig. 6).

It should be noted that thermal cycling forms part of the dynamic tests to be described later.

Table 1. *Change of potential measured against a saturated calomel electrode of ten MnO₂-electrodes kept in a thermostated bath of saturated Ca(OH)₂. The temperature was cycled from room temperature to 70°C, kept for 4 days and returned to room temperature*

Date Time Temp.	16-05-94 1300 h 23°C	16-05-94 1350 h 70°C	17-05-94 1200 h 70°C	17-05-94 1500 h 70°C	18-05-94 1500 h 70°C	20-05-94 0830 h 70°C	20-05-94 1130 h 40°C	20-05-94 1430 h 25°C	25-05-94 0900 h 21°C
	mV	mV	mV	mV	mV	mV	mV	mV	mV
1	168	147	149	141	129	130	162	169	163
2	168	149	150	143	131	129	158	168	172
3	168	147	144	140	129	131	159	167	173
4	169	148	147	141	129	131	163	168	174
5	170	149	150	145	133	133	160	169	173
6	167	144	146	142	130	130	156	165	170
7	168	145	148	146	132	131	159	169	173
8	165	146	148	146	133	132	160	169	173
9	164	144	141	141	132	132	160	169	173
10	167	144	148	140	132	132	160	169	173

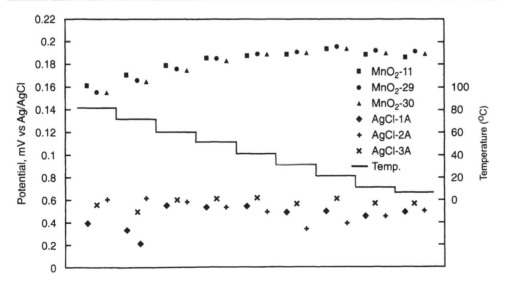

Fig. 6 *Change of potential with temperature for three MnO₂-electrodes and three Ag/AgCl-electrodes. Each temperature was kept constant for 4 days.*

2.6. Dynamic tests

A special dynamic test was designed, in which the electrodes were close to cathodically protected steel in concrete, while the concrete blocks went through climatic variations.

A total of six blocks with dimensions and electrode arrangements as shown in Fig. 7 were cast from concrete with 325 kgm^{-3} Portland cement and a w/c ratio of 0.6. Two of the blocks contained 2% (by weight of cement) of chloride added as CaCl$_2$. These two blocks also contained Ag/AgCl-electrodes of the types mentioned earlier. The impressed current cathodic protection was with a constant current density of 80 mAm^{-2} of the steel surface areas. Of the four blocks with chloride-free concrete, two were exposed weekly to a NaCl-solution.

The six blocks thus formed two sets of three blocks with three different levels of chloride exposure. One set of blocks was exposed to cycles of 12 h in a 40°C condensing atmosphere alternating with 12 h drying in the laboratory air. The other set went through 80 hour cycles of temperatures shifting from –10°C to +80°C.

During exposure the 'On'-potentials were measured in 4-hour intervals. Once per week the 'Off'-potential of the steel was measured relative to all embedded electrodes and a surface held Cu/CuSO$_4$-electrode, one second after switching off the current. The measurements were made simultaneously using three or four voltmeters as required. A separate measurement of the potential difference between the two embedded MnO$_2$-electrodes was also made.

The 'On'-potentials are affected by strong IR-drops and need not be discussed. The 'Off' potentials are indicative of the level of CP (cathodic protection) achieved, but are also sensitive to the exact timing of the measurement, because the potentials are still changing 1 s after interrupting the current. The curves for all six test conditions are shown in Fig. 8(a–f). Some of the early measurements were unreliable due to technical difficulties and have been omitted.

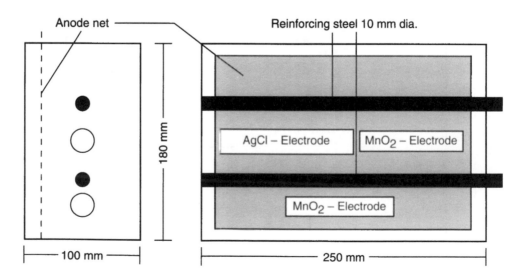

Fig. 7 *Drawing of test block used for dynamic tests — see text for further explanation.*

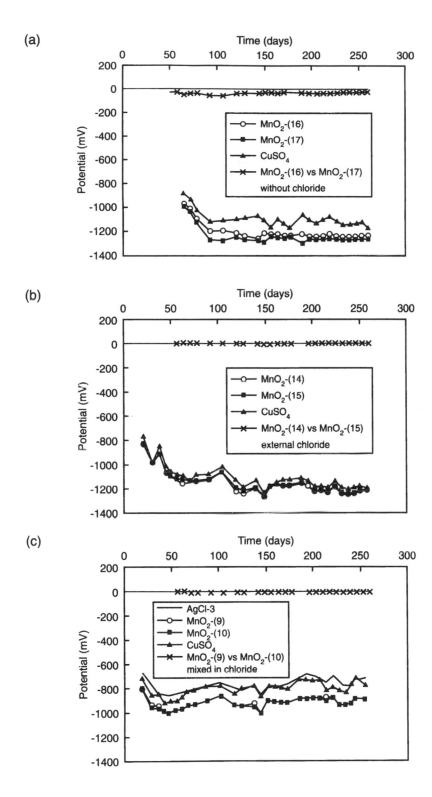

Fig. 8 (a–c) Results of dynamic tests for test blocks cycled between 40 °C condensing atmosphere and laboratory air. See text for further explanation.

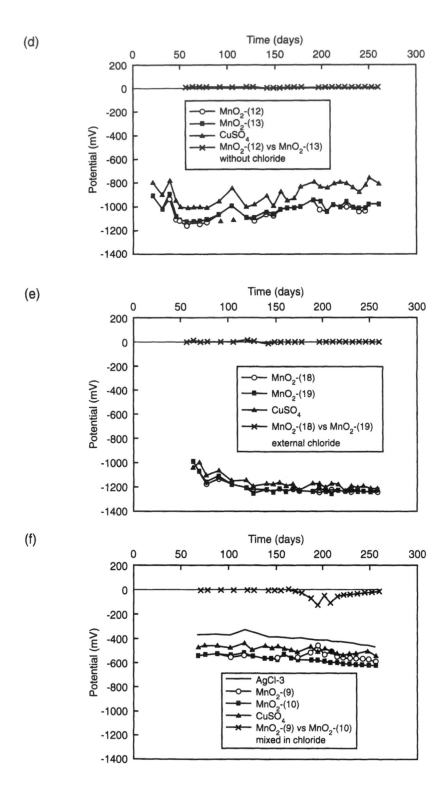

Fig. 8 (d–f) *Results of dynamic tests for test blocks cycled from −10°C to +80°C. See text for further explanation.*

The curves show that the pairs of installed MnO_2-electrodes are in good mutual agreement, except in Fig. 8(a), where there is a 30–60 mV difference, which cannot be explained, and Fig. 8(f), where the results seem to be influenced by localised corrosion which has occurred in spite of the CP under the rather extreme experimental conditions. Note that the general level of the off-potentials is higher than in the other curves. Note also that one of the MnO_2-electrodes is placed very near to the Ag/AgCl-electrode and stays in agreement with this, while the other is farther away and apparently picks up the IR-drop from a local anode.

3. Field Experience

The MnO_2-electrode has found wide acceptance in industry, and thousands of electrodes are installed in bridges, tunnels, docks and swimming pools in Europe, in the Middle and Far East and elsewhere. Most are sensors in potentiostatically controlled current supplies for CP installations and some are part of monitoring systems. A few reported failures seem to have been of a mechanical nature, probably the result of damaged or broken electrical cables.

Following requests from the manufacturer, some users have reported attempts to calibrate embedded electrodes against surface-held reference electrodes, but it is in practice impossible to calibrate single embedded electrodes against surface-held electrodes, due to the great variability of the surface junction potential. In the few cases, where MnO_2-electrodes have been installed in pairs, they have shown parallel behaviour [4].

Results from a large-scale test of embeddable electrodes and electrochemical monitoring techniques used on Gimsøystraumen Bridge in Norway have recently been published [7,9]. Where electrodes were installed in pairs, MnO_2-electrodes were in better agreement than graphite and lead electrodes. Of the installed 64 electrodes, 25% failed or showed instability after 3 years. Some, out of a total of 32, MnO_2-electrodes were among them, but the full statistics were not given. The electrodes were compared with external, surface held electrodes which only emphasised the variability of the junction potentials. Unfortunately no measurements were made with external electrodes touching the concrete in drilled holes, using saturated KCl or KNO_3 as the contacting electrolyte. This technique was earlier shown to give the most stable and representative control measurements of embedded electrodes [2].

MnO_2-electrodes were chosen for two interesting applications on the Gimsøystraumen Bridge, namely for polarisation resistance measurements and for determination of oxygen limiting currents on electrodes held at a constant cathodic potential. Obviously great stability of the chosen reference electrode is an important requirement in this case, and the clear trends in the measurements of the effect that surface treatments has on oxygen access [9] has demonstrated the suitability of the MnO_2-electrode for this application.

4. Discussion

As fabricated, MnO_2-electrodes exhibit slightly different calibration potentials within a reasonably narrow scatterband of ±10 mV. Individual electrodes have maintained their relative potential differences with even less variation under a variety of climatic conditions in ongoing tests for periods of up to 4 years so far. The exceptionally good short term stability, in the order of 10 μVh^{-1}, make them ideally suited for electrochemical noise measurements and Electrochemical Impedance Spectroscopy, where such measurements are considered.

The polarisation behaviour is characterised by an internal resistance of 1–2 kΩ and an ability to deliver (or receive) small currents for prolonged times. The change in potential after such polarisation was found to be less than 1 mV per coulomb.

Because of the very good properties mentioned above and also the ability to withstand both low and high temperatures (tests have been conducted in the range from –10°C to 80°C) manganese dioxide reference electrodes have found wide application in the field, where their excellent long term stability has been proved.

The latest example of the use of MnO_2 electrodes in measurements of oxygen limiting current supplement the previously known applications for monitoring of CP installations and reinforcement potential.

However, recently published results show that a lot of unexplained problems occur when measurements made using embeddable reference electrode are interpreted.

The most important stumbling block for the interpretation of potentials of metals in concrete is the existence of strong junction potentials at surfaces and probably also internally in the concrete. This problem has been discussed by Arup *et al.* [2] and by Bennet *et al.* [3]. Myrdal [6] has discussed the diffusion potentials at liquid junctions and compared calculations with experimental data. His experiments were limited to liquids making contact through opening valves and joining wet sponges. The contact between an electrolyte and a cement-based capillary system has added problems stemming from the double-layer effects in the fine pores, especially in semi-dry concrete. Two of the authors are presently engaged in work on this subject, and some results are published in another paper in this present volume [10].

References

1. H. Arup, Embeddable Reference Electrodes for Use in Concrete, Nordic Concrete Research 1990. Publication No. 9, p.6–13.

2. H. Arup and B. Sørensen, A new embeddable reference electrode for use in concrete. *Corrosion '92*, Paper 208. NACE, Houston, Tx, 1992.

3. J. E. Bennet and T. A. Mitchel, Reference electrodes for use with reinforced concrete structures. *Corrosion '92*. Paper 91. NACE, Houston, Tx, 1992.

4. P. Castro, A. A. Sagüés, E. I. Moreno, L. Maldonado and J. Genescá, Characterisation of activated titanium solid reference electrodes for corrosion testing of steel in concrete, *Corrosion*, 1996, **52** (8), 609–617.

5. S. Atlung and T. Sørensen, The potential of battery active manganese dioxide, *Electrochim. Acta*, **26**, (10), 1447–1456.

6. R. Myrdal, Phenomena that disturb the measurement of potentials in concrete, *Corrosion '96*. Paper 339. NACE, Houston, Tx, 1996.

7. K. Videm and R. Myrdal, *Proc. Int. Conf. on Repair of Concrete Structures*, Svolvær, Norway, May 1997, 375–390.

8. H. Arup, O. Klinghoffer and J. Mietz, Long term performance of MnO_2 reference electrodes in concrete, *Corrosion '97*. Paper 243. NACE, Houston, Tx, 1997.

9. Ø. Vennesland, *Proc. Int. Conf. on Repair of Concrete Structures*, Svolvaer, Norway, May 1997, 253–262.

10. H. Arup and O. Klinghoffer, Junction potentials at a concrete/electrolyte interface, this volume, p. 31–39.

6

Corrosion Inhibitors for Steel in Concrete

B. ELSENER, M. BÜCHLER and H. BÖHNI

Institute of Materials Chemistry and Corrosion, Swiss Federal Institute of Technology, ETH Hönggerberg,
CH-8093 Zürich, Switzerland

ABSTRACT

Inhibitors, chemical substances that prevent or retard corrosion, are applied to new and existing structures where corrosion has to be prevented. For repair work, inhibitors can be present in paints for the reinforcement, added to repair mortars or applied from the surface. A survey on literature results on inhibitors for steel in concrete is given; in particular calcium nitrite, the migrating inhibitors (MCI or SIKA)™ and MFP (monofluoro phosphate) are addressed. Results from a two year laboratory study on corrosion of steel in presence of migrating corrosion inhibitors (MCI) are presented. The problem of testing the efficiency of inhibitors for steel in concrete in the laboratory and on-site is discussed. A critical evaluation of corrosion inhibitors to be used in reinforced concrete structures with respect to concentration dependence, durability and measurement and control of the inhibitor action is given.

1. Introduction

In general, reinforced concrete has proved to be successful in terms of both structural performance and durability. However, there are instances of premature failure of reinforced concrete components as a result of corrosion of the reinforcement (rebar). The two factors provoking corrosion are the ingress of chloride ions from de-icing salts or sea water or the reaction of the alkaline pore solution with carbon dioxide from the atmosphere, a process known as carbonation. From the point of view of corrosion protection of the rebars two different situations have to be distinguished, i.e. new and existing structures:

- with *new structures*, the most effective measure for durability can be achieved in the design stage by using adequate concrete cover and high concrete quality. This will prevent aggressive substances, e.g. chloride ions from deicing salts or sea water to reach the rebars within the design life. Additional protective measures can be applied, such as admixtures to concrete in order to decrease its permeability, the use of more corrosion resistant materials for the reinforcement (e.g. stainless steels*), electrochemical protection systems (e.g. preventive cathodic protection) or others.

---

* A State of the Art Report on Stainless Steel in Concrete has been published as Number 18 in the EFC Series.

- with *existing structures* the deterioration process may have reached different stages according to age, exposure condition, concrete cover and quality: for a corrosion risk situation or at the onset of corrosion, preventive measures may be applied, whereas in severely corroding structures repairs have to be carried out.

Inhibitors, i.e. chemical substances that prevent or retard corrosion, are proposed (and used) for new structures and as preventive and as repair measures for existing reinforced concrete structures. The method of application varies: in new structures inhibitors are admixed in sufficiently high concentrations to the fresh concrete; on existing structures where the onset of corrosion has to be prevented, inhibitors are applied at the concrete surface; for repair work inhibitors can be present in paints for the reinforcement or in repair mortars. The present paper presents a literature survey on inhibitors for steel in concrete with particular attention given to calcium nitrite (DCI)™, the migrating (see Section 3 below) corrosion inhibitors (SIKA or MCI) and MFP (monofluorophosphate) are addressed. Results from a two-year laboratory study on corrosion of steel in concrete in presence of the migrating corrosion inhibitor (MCI) are presented. The problem of testing different inhibitors for steel in concrete is addressed and — as far as available — results from field tests with inhibitors are presented. Finally, a critical evaluation of corrosion inhibitors for steel in concrete is given.

2. Corrosion Inhibitors for Use in Concrete

Corrosion inhibitors are chemical substances that, when added in adequate (preferably small) amounts to concrete, can prevent or retard corrosion of steel in the concrete. They should not show adverse effects on the concrete properties (e.g. compressive strength) and not adversely affect the nature and microstructure of the hydration products. Several admixtures — pure compounds or mixtures — have been available as corrosion inhibitors for a long time and are claimed to offer protection for reinforcing steel in concrete against chloride-induced corrosion. These admixtures are used as a preventive measure and are added to fresh concrete or to repair products such as paints for reinforcing steel, adhesion bridges and mortar. More recently, new interest has arisen on the use of inhibitors as rehabilitation or curative measures. In this approach the compounds are applied onto the concrete surface and should penetrate through the concrete to the steel to stop or retard corrosion. However, there are conflicting opinions about the effectiveness of these compounds in corrosion protection.

3. Mechanism

The mechanistic action of corrosion inhibitors has to be considered in the context of whether uniform or pitting corrosion prevails. The overwhelming majority of literature on corrosion inhibitors deals with the effects of inhibitors on uniform corrosion. Thus, inhibitors can be classified into [1,2]:

- adsorption inhibitors, acting specifically on the anodic or on the cathodic partial reaction of the corrosion process or on both reactions,

- film forming inhibitors blocking the surface more or less completely, and

- passivators favouring the passivation reaction of the steel (e.g. hydroxyl ions).

Inhibitors for pitting corrosion are by far less studied [3]. Chloride ions are usually responsible for pitting corrosion, with the electrochemical pitting potential $E_{pit} = C - B \log(a_{Cl^-})$ in which C and B are constants and a_{Cl^-} is the chloride activity (concentration). Inhibitors for pitting corrosion can act by

- a competitive surface adsorption process of inhibitor and chloride ion (reducing the effective chloride content on the passive surface),

- buffering of the pH in the local pit environment, and

- competitive migration of inhibitor and chloride ions into the pit so that the low pH and high chloride contents necessary to sustain pit growth cannot develop.

Migratory corrosion inhibitors, derived from vapour phase (VPI) or volatile corrosion inhibitors (VCI), have the property of being transported through the gas phase to the metal surface and can be of any of the above types. Commercial inhibitors frequently are blends of several compounds, so that more than one mechanism may be operating.

4. Inhibitors as Repair Strategy

Inhibitors are one of several possible repair strategies [4, 5] and their use on reinforced concrete structures has to be planned with the same care as the construction of new structures. Before any decision to use inhibitors as a rehabilitation method is taken, the following analysis of the situation is recommended in order to achieve cost effective and durable repairs [4]. The following factors need to be taken into account.

1. *Structural condition:* A thorough condition assessment of the structure (or part of it) should include a visual survey, the identification of structural cracks, deformations etc. in order to clarify if, and to what extent, structural repairs have to be carried out.

2. *Cause of deterioration:* Any condition assessment should be of sufficient duration to identify clearly the cause(s) of the observed degradation. It is useful to start with non-destructive techniques (e.g. potential mapping [6] to locate corroding zones) before applying destructive techniques (e.g. core drilling for chloride analysis).

3. *Expected service life of the structure:* The owner of the structure has to decide the future use of the structure and to define the desired service life.

Rehabilitation with inhibitors has the advantage that only a minimum intervention is required, some local repairs may be necessary because of the presence of cracks, spalling etc. or for aesthetic reasons. To effectively prevent or stop corrosion, the inhibitor has to be present in a sufficiently high concentration in the region of the reinforcement.

5. Short Literature Review

The primary 'inhibitor' for steel in concrete is the hydroxide ion (OH^-) which when present at a sufficiently high concentration will promote the formation of a stable oxide/hydroxide film (passive film) at the steel surface. While numerous other inhibitors have been suggested, only a small group has been seriously studied. Literature reviews have been published [7,8]. Previous studies on chemical substances that prevent the onset of pitting corrosion have focused mainly on anodic corrosion inhibitors, especially calcium nitrite, sodium nitrite, stannous chloride, sodium benzoate and some other sodium and potassium salts (e.g. chromates).

5.1. Nitrites

The efficiency of calcium and sodium nitrite as inhibitors in concrete has been reported by several authors [9,10] since the early 1970s. The investigations, which have been conducted with different experimental techniques in solutions, in mortar and in concrete, revealed a critical concentration ratio between inhibitor (nitrite) and chloride of about 0.6. This implies that quite high nitrite concentrations have to be present in the pore water of the concrete to act against chlorides penetrating from the concrete surface. Nitrite acts as a passivator due to its oxidising properties and stabilises the passive film according to the reaction [10]:

$$2\,Fe^{2+} + 2\,OH^- + 2\,NO_2^- \rightarrow 2\,NO + Fe_2O_3 + H_2O \tag{1}$$

The effect of nitrite in enhancing passivity is related to its ability to oxidise ferrous to ferric ions, which are insoluble in aqueous alkaline solutions. Sodium nitrite, despite its good inhibitive properties, has caused severe strength loss when admixed to concrete. Calcium nitrite acts as an accelerator, and normally requires the addition of a water reducer and retarder in the concrete mixture. Calcium nitrite was included in the tests in a comparative study to evaluate corrosion inhibitors [11,12]. As a result of an error in the recommended dosage the concentration was too low and nitrite was effective only for specimens with low chloride concentration. This confirms the necessity to reach a critical concentration ratio for effective corrosion protection. Concentrations of nitrites that are too low may cause the risk of an increased corrosion rate — as has been found in laboratory studies on cracked reinforcing beams [3]. Nitrites — when applied according to the appropriate specifications together with high quality concrete — have a long and proven track record in the USA, Japan and in the Middle East [13]. However, because of environmental regulations and concern about a possible increase of the corrosion rate when the inhibitor is present in insufficient concentrations [3] they have found only few applications in Europe.

5.2. Stannous Chloride

Stannous chloride (SnCl$_2$. 2H$_2$O) was tested as an inhibitor of chloride-induced corrosion of steel in alkaline solutions [14]. The results showed that stannous chloride did not dissolve well in the pore solution (lime water) and could not protect the steel completely from corrosion. Stannous chloride appeared to be a non-corrosive accelerator that causes concrete to harden faster. Due to the chlorides dissolved from the 'inhibitor' this substance is not used.

5.3. Sodium Benzoate

Sodium benzoate resulted in a decrease in compressive strength when admixed to fresh concrete, the inhibitive properties were moderate [15].

More recently, several investigations on the evaluation of corrosion inhibitors for the rehabilitation of reinforced concrete structures have been conducted. The inhibitors studied were surface applied liquids and concrete admixtures.

5.4. MFP (Sodium Monofluoro Phosphate, Na$_2$PO$_3$F)

MFP (sodium monofluoro phosphate, Na$_2$PO$_3$F) has been studied — as described in a recently published paper [16] — in the laboratory as an inhibitor in neutral aqueous solutions, then as an admixture to deicing salts and has been patented in the U.S. based on the results of Canadian studies (DOMTAR Inc.). MFP cannot be used as an admixture due to the chemical reaction with the fresh concrete, and so it has to penetrate from the concrete surface to the steel. Its mechanism and action as inhibitor against chloride-induced corrosion has been studied in the laboratory [17]. Solutions of saturated Ca(OH)$_2$ and small mortar samples were used for the tests. The main result from the experiments [17] was stated to be that a critical concentration ratio of MFP / chlorides greater than 1 has to be achieved, otherwise the reduction in corrosion rate is not significant. No complete repassivation was found after the onset of chloride induced corrosion. MFP was tested as a preventive inhibitor by applying several flushings before the ingress of chlorides [17]; it stopped corrosion during the test duration of 90 days even at chloride concentrations as high as 2% by weight of cement. In solutions containing Ca(OH)$_2$ MFP reacts with the calcium ion to form the insoluble products calcium phosphate and calcium fluoride [16,17], thus the active substance, the PO$_3$F$^-$ ion, disappears from the pore solution. The inhibitive effect of MFP for corroding reinforcement in carbonated concrete has been studied [18]. Repeated MFP treatments with dry and immersion cycles have been found to provide a suitable method to allow the penetration of the inhibitor to the steel, although high concentrations and long treatments are needed to reduce active corrosion significantly.

The main problem using MFP as a surface applied liquid is the penetration to the reinforcement since this is where the inhibitive action is required. In early field tests in Switzerland no sufficient penetration of MFP was found [19,20]. This was partly caused by the concrete being too dense — the cover depth being greater than 45 mm — or an insufficient number of MFP applications to the surface. Heating of the concrete surface did not show any beneficial effect for MFP penetration [20]. In more recent field applications [16], e.g. on the Peney Bridge near Geneva [21], concrete buildings and balconies, MFP was applied to cleaned, dry concrete surfaces in up to 10 passes and the concrete became impregnated to the reinforcement level in a few days or weeks.

5.5. Alkanolamines and Amines

Alkanolamines and amines and their salts with organic and inorganic acids have been described and patented for different applications, such as for the protection of steel in cementitious matrices [22 and literature cited therein]. A European Patent Application published in 1987 [23] describes the use of one or more hydroxyalkylamines having molecular weights ranging from about 48 to 500 and vapour pressures at 20°C ranging from 10^{-4} to 10 mm Hg that are employed as the major ingredient of a corrosion inhibitor to be mixed into hydraulic cement slurry. The hydroxyalkylamines provide corrosion protection to iron and steel reinforcing members embedded in concrete and do not substantially affect the air entraining capacity. Typical compounds mentioned include diethanolamine, dimethylpropanolamine, monoethanolamine and dimethylethanolamine. Compressive strength and time of setting are not altered by more than 20%. Figure 1 shows the corrosion rates obtained in laboratory experiments in presence of different hydroxyalkylamines or their mixtures compared to the control sample. Another patent specification [24] describing corrosion inhibition in reinforced concrete addresses

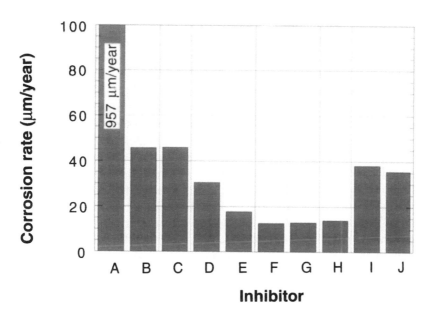

Fig. 1 *Corrosion rate [μm/year] of steel in concrete with different inhibitors mixed in. A: Control; B: dimethylaminoethoxyethanol; C: N,N,N'-trimethyl (hydroxyethyl) ethylenediamine); D: N,N,N'-trimethyl (hydroxyethyl)-1,3-propane diamine; E: N,N,N'-trimethyl (hydroxypropyl)1,3-propane diamine; F: methyldiethanolamine; G: triethanolamine; H: monoethanolamine; I: dimethylethanolamine; J: dicyclohexylamine.*

vapour phase corrosion inhibitors (VPI) or volatile corrosion inhibitors (VCI). As preferred inhibitors for reinforcement dicyclohexylamine nitrite (DCHN), cyclohexylamine benzoate (CHAB) and cyclohexylamine carbamate (CHC) or mixtures of CHAB and CHA are proposed. The aim in mixing is to get a VPI with a fast initial release and one with a slow vaporisation. It is proposed to bring the VPI into the concrete by drilling holes at suitable places. The spacing of the drill holes depends on the amount of reinforcement, the volatility of the VPI, the porosity of the concrete and the VPI content in the holes. Only short term corrosion tests and no evidence of long term efficiency are given [24]. Further information on vapour phase inhibitors can be found in a review [25] and research papers [26]. It is interesting to note that the patent application explicitly reports the advantage of all the hydroxyalkylamine compounds being water soluble, so that they demonstrate mobility within concrete structures when water is applied. It is claimed that these inhibitors can be applied to existing reinforced concrete structures and that the corrosion inhibitor will be carried by water into the proximity of the reinforcing steel. The inhibitor could thus be included in hydraulic cement overlays on old concrete structures. Some vapour phase migration of the inhibitors is believed to occur as well as a result of the vapour pressure of the inhibitor.

Several proprietary blends from different producers (as e.g. Cortec VCI-1337™ or MCI-2020™, Cortec VCI-1609™ or MCI-2000™, SIKA Armatec 2000™ or SIKA Ferrogard™) are based on the principle of using alkanolamines and amines and their salts with organic and inorganic acids. The Cortec inhibitors were included in the comparative tests of the SHRP (Strategic Highway Research Programme) project [11,12] and good inhibitive properties were reported: the corrosion rate (determined by linear polarisation resistance) decreased and the corrosion potential shifted to more positive values. The inhibitors penetrated from the repair material through to the next layer of steel reinforcement in the parent concrete [11]. A recent comparative test of different organic amines in alkaline solutions [27] showed very good corrosion inhibition of the commercial MCI inhibitor, pure dimethylethanolamine itself being practically ineffective. SIKA Ferrogard 901 was tested as an admixture in mortar and concrete samples exposed to chlorides [28]. After one year of test corrosion had started in specimens with w/c (water:cement ratio) = 0.6, the chloride threshold values were in all cases higher for the inhibitor containing samples (4 – 6% Cl^- by weight of cement) compared to the control samples (1 – 3% Cl^-). Diffusion experiments [22,27,28] showed that these types of inhibitors can migrate through the concrete although great discrepancies in the measured diffusion rates exist. This might partially be accounted for by the different experimental set-up and measuring techniques used. Mechanistically these inhibitors are claimed to belong to the film forming inhibitors, thus blocking both the anodic and the cathodic reaction of the corrosion process. Film formation has indeed been revealed by analysing the steel surface after immersion in inhibitor-containing solutions by surface analytical techniques [29,30].

5.6. Organic Based Admixtures

Organic based admixtures were proposed in a United States Patent [31]. An admixture is marketed as Rheocrete from Master Builders Inc. The admixture comprises an oil/water emulsion, in which the oil phase comprises an unsaturated fatty acid ester of an aliphatic carboxylic acid with a mono-, di- or trihydric alcohol and the water phase comprises a saturated fatty acid, an amphoteric compound, a glycol and a soap. The admixture is added to concrete prior to placement. Upon contact with the high pH environment of the concrete the emulsion collapses allowing contact between the active agents and the steel reinforcing bars. Anodic polarisation and time to corrosion tests showed good inhibitor efficiency. The mechanism is described as formation of a physical barrier against the action of aggressive agents such as chloride ions [32]. In addition, the chloride ingress into concrete is reduced.

6. Laboratory Study with Migrating Corrosion Inhibitors

Migrating corrosion inhibitors (MCI 2000 and 2020) have been tested in a two years laboratory study in solution and in mortar samples [33] as inhibitors against carbonation or chloride induced corrosion.

6.1. Experiments in Solution

In a first series of experiments the critical concentration of MCI 2000 was determined. The samples were immersed for 7 days in chloride free saturated Ca(OH)$_2$ solution with different contents of inhibitor, the solution was then changed to saturated Ca(OH)$_2$ with 1M NaCl. As shown in Fig. 2 the electrode potentials remained in the passive state only with addition of 10% inhibitor.

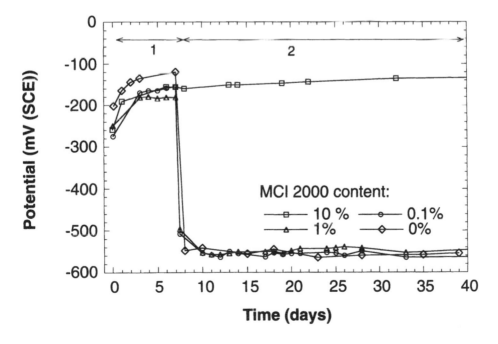

Fig. 2 *Effect of MCI 2000 inhibitor concentration on open circuit potential of mild steel in satd. Ca(OH)$_2$ solutions with 1M NaCl [33]. 1: prepassivation in satd. Ca(OH)$_2$; 2: immersion in satd. Ca(OH)$_2$ + 1M NaCl, open to air.*

Chemical analysis revealed that the inhibitor blend consisted of two main components, a volatile amine (about 95%) and non-volatile part (5%) that were separated by distillation at 30°C. Electrochemical tests were conducted with the two main components in alkaline solutions. Potential and linear polarisation resistance (LPR) measurements showed that both components of the inhibitor, the volatile and the non-volatile, present alone in solution could not prevent initiation of corrosion (Fig. 3). Polarisation resistance values measured in solutions with either of the inhibitor components indicated a reduction in the corrosion rate by a factor of about 2–3 compared to solutions without inhibitor.

Electrochemical impedance spectroscopy (EIS) measurements conducted after immersion of steel in saturated $Ca(OH)_2$ solutions without and with 10% of inhibitor clearly revealed a second time constant at high frequencies on the sample immersed in solution with inhibitor (Fig. 4). This indicates some film formation on the passive steel surface in presence of the inhibitor MCI 2000.

6.2. Surface Analysis

The interaction of the inhibitor MCI 2000 and of pure 2-dimethylethanolamine solvent with passive iron surfaces has been studied with X-ray photoelectron spectroscopy (XPS) and with ToF-SIMS (Time of Flight Specific Ion Mass Spectroscopy) [34]. Mirror like polished samples were immersed in alkaline solutions simulating the pore solution of concrete for one hour, one day and three days. The solutions studied were blank (uninhibited), with 10% of DMEA and with 10% of inhibitor. After removal

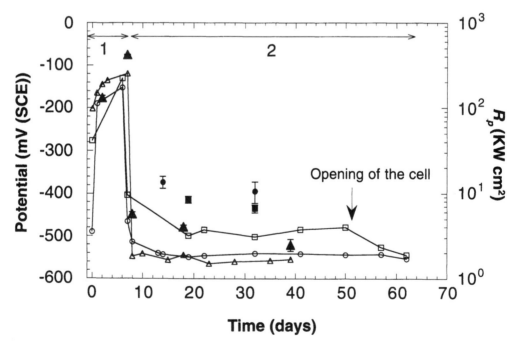

Fig. 3 Open circuit potential (Δ □ ○) and polarisation resistance (▲ ■ ●) of mild steel in satd. $Ca(OH)_2$ solutions with 1M NaCl without inhibitor (Δ ▲), with 10% of volatile compound (□ ■) and with 10% of the non-volatile compound of the inhibitor (○ ●)[33].

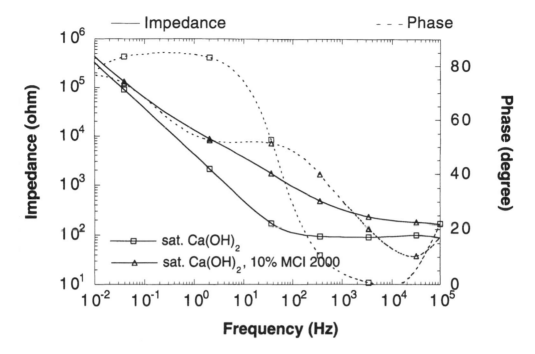

Fig. 4 *Impedance spectra (Bode plot) for steel immersed in satd. Ca(OH)$_2$ solution without and with 10% of inhibitor MCI 2000. Immersion 3 h [33].*

from the test solutions and rinsing with distilled water, all samples were mirror like without deposits or precipitates. Samples from solutions with 10% of inhibitor blend were hydrophobic, thus the water drops did not wet the surfaces. From the quantitative analysis of the XPS data a thickness of the organic layer on the surface after 1 h of immersion in the solution with inhibitor of 6.3 ± 0.2 nm (density assumed 1) was calculated whereas only 3.8 ± 0.2 nm was measured in the blank and in DMEA solutions. This indicates a specific adsorption of the inhibitor blend on the surface as has been found by EIS measurements (Fig. 4). Further immersion for 24 or 96 h revealed no significant increase in film thickness.

Highly surface sensitive and molecular fragment specific ToF-SIMS measurements revealed the presence of two prominent lines at negative m/z (mass/charge) fragments of 121 and 281, both after deposition of the inhibitor blend on gold and after immersion in alkaline solution with 10% of inhibitor on iron [34], thus the inhibitor blend (as mentioned in the patent applications) contains other chemical substances (e.g. benzoates).

6.3. Mortar Experiments

The inhibitor efficiency against chloride induced corrosion was studied using mortar samples (lollipops) with w/c ratio of 0.5 and inhibitor contents of 0, 0.015, 0.075 and 0.375 % by weight of mortar. After curing for 70 days in 100% R.H. all samples (6 samples for each inhibitor concentration) were exposed to cycles of 1 day immersion in 6% NaCl and 2.5 h drying in air, to initiate chloride induced corrosion. The time to corrosion initiation in presence of the inhibitor was increased (Fig. 5). The first sample of the series with the highest concentration of inhibitor started to corrode after 90 instead of 50 days. On the other hand, no significant reduction in the corrosion rate of samples after the initiation of corrosion was found. A reduction in corrosion rate on already corroding rebars by inhibitor penetration was hardly apparent. It was shown that the volatile component of the inhibitor was evaporating from the mortar [33].

6.4. Field Tests

Despite several site projects and applications no conclusive field tests have so far been reported in the literature. Some ongoing field trials lack well defined condition assessment prior to the inhibitor application and, on others, corroding areas were repaired conventionally prior to the inhibitor application. In Switzerland a comparative field test with surface applied inhibitors (including MFP and SIKA Ferrogard) on a chloride contaminated side wall has been started with a thorough characterisation of the test sites. First results will not be obtained before the end of 1998.

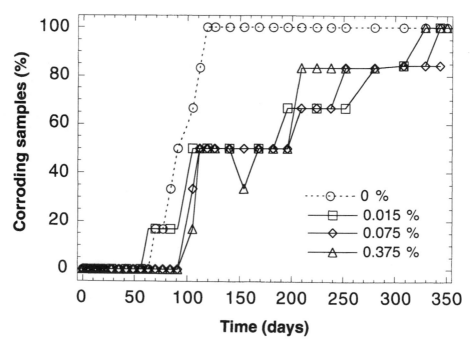

Fig. 5 Percentage of actively corroding mortar samples during cyclic immersion tests with 2.5 h drying and 1 day immersion in 6% NaCl solution [33]. Each series with different inhibitor content consisted of six samples.

7. Critical Evaluation of Corrosion Inhibitors

Corrosion inhibitors in new reinforced concrete structures, in concrete repair systems or as surface applied liquids should prevent or at least delay the depassivation of the steel and / or reduce the corrosion rate of steel in concrete. Several basic conditions have to be fulfilled for an efficient and durable inhibitor action:

1. The inhibitor should not adversely affect concrete properties (strength, freeze-thaw resistance, porosity etc.) and should be environmental friendly.

2. The inhibitor has to be present at the reinforcing steel at a sufficiently high concentration with respect to the aggressive (chloride) ions.

3. The inhibitor concentration should be maintained over a long period of time.

4. The inhibitor action on corrosion of steel in concrete should be measurable.

7.1. Concrete Properties — Environment

Adverse effects of inhibitors on concrete properties can be tested in the laboratory. MFP strongly retards concrete setting and thus can not be used as admixture, on the other hand it has been reported to reduce freeze thaw attack [16,19]. Calcium nitrite is an accelerator for concrete setting. Aspects of environmental compatibility of all chemicals used in the construction industry become more and more important both from a legal and financial point of view. Harmful or toxic inhibitors can hardly be used in Europe because of severe environmental protection regulations. The amine-based inhibitors are alkaline and should not contaminate water or earth. More and more concerns are being raised regarding the use of all types of concrete admixtures because concrete, when demolished at the end of the service life of a structure, may be considered as hazardous waste with very high cost for disposal.

8.2. Concentration Dependence

The available literature reports a concentration dependent effect of inhibitors. A critical inhibitor / chloride ratio has to be exceeded. For new structures the inhibitor dosage thus has to be specified with respect to the expected chloride level over the design life of the structure. To be on the safe side, a certain overdosage is necessary. For calcium nitrite the ratio of Cl^-/NO_2 should be in the range of 1.5 to maximum 2 [10], for MFP a molar ratio of 1:1 has been reported [17] and also in this work high inhibitor concentrations are reported to be necessary to prevent the onset of corrosion (Fig. 2) [33].

Surface applied inhibitors on existing structures may present even more difficulties in achieving the necessary concentration at the rebar level as has been found, e.g. in field tests with MFP [20]. First, because chloride contamination or carbonation may vary strongly along the surface (e.g. as found on a typical sidewall exposed to splash water [35]), secondly, because the cover and permeability of the concrete may vary as well and thirdly, because the inhibitor may react with pore solution components. In the application notes of surface applied inhibitors not only an average weight of inhibitor solution to be applied per square metre of concrete should be specified (or the number of applications) but also the critical concentration to be achieved at the rebar level. This is usually omitted, in part due to the lack of analytical methods that are available to measure the inhibitor concentration but also because inhibitors may be washed out from the concrete or evaporate.

7.3. Measurement and Control of Inhibitor Action

One of the main difficulties in evaluating the performance of inhibitors is to assess the inhibitor action on rebar corrosion 'on site'. The interpretation of half cell potential measurements may present difficulties due to changes in the concrete resistivity [20,36,37]. Further, a reduction of corrosion rate due to an inhibitor action may not be reflected in a straightforward manner in the half-cell potentials since these may become more negative or more positive after inhibitor application, depending on the mechanism of the inhibitor action. Shifts in the half-cell potential may occur also due to the wetting and drying of the concrete [20,36]. Results of corrosion rate measurements on-site depend on the type of device used for the measurements and so far can be interpreted only by specialists [6]. The main problems are the daily and seasonal changes of the corrosion rate with temperature and concrete humidity making it difficult to evaluate inhibitor action. Small scale field trials using two proprietary vapour phase inhibitors (VPI) report a modest but statistically significant reduction in corrosion rate [38].

7.4. Durability of the Inhibitor Action

On concrete structures exposed to splash water or rain the inhibitors may be washed out at least at the concrete surface and the critical concentration ratio may not be maintained over time. Calcium nitrite as a passivating inhibitor reacts with iron(II) ions to form a protective film. Repetitive pit initiation processes can lead to a reduction in the nitrite content near the rebar and the protective action may be lost. Increase in pitting depth has been reported in cracked concrete after 2.5 years of exposure [3,39]. Admixed or surface applied inhibitors that are claimed to migrate (also in the gas phase) through the concrete can evaporate again and the necessary concentration for protection may not be reached after long time — or, more probably, the inhibitor application has to be repeated after several years.

8. Concluding Remarks

Inhibitor technology could in principle offer a cost effective way for the rehabilitation of reinforced concrete structures. Several inhibitors have been tested in the laboratory and are commercially available today both as admixtures for fresh mortar or concrete or as surface applied liquids.

In some contrast to the marketing publicity that shows very promising laboratory data, only very few documented results from field tests or field applications are available. Usually the use of inhibitors as admixtures is recommended only for high quality concrete (w/c ratio < 0.45), on structures with high concrete cover and good workmanship. The use of inhibitors as surface applied liquids can be recommended with the condition that the chloride content in the concrete is not too high and that all corroding, cracked or spalled areas are repaired prior to the inhibitor application. Thus, inhibitors for reinforced concrete may prolong service life of new and/or repaired structures — but both conclusive field tests and long term experience are still missing.

If corrosion inhibitors are to be used effectively as a repair strategy, it is important:

- to specify the concentrations that are needed at the reinforcement level;

- to propose and develop suitable means for demonstrating that such conditions are actually achieved and maintained for long times taking into account diffusion, leaching etc; and

- to demonstrate on-site that inhibitors prevent or at least delay the onset of corrosion or reduce the corrosion rate.

References

1. G. Trabanelli, Corrosion inhibitors, in *Corrosion Mechanisms*, Ed. F. Mansfeld. Marcel Dekker, N.Y., 1986, chapter 3.
2. U. Nürnberger, Corrosion inhibitors for steel in concrete, *Otto Graf J.*, 1996, **7**, 128.
3. D. W. DeBerry, Organic Inhibitors for pitting corrosion, in *Review on Corrosion Inhibitor Science and Technology*, Eds A. Raman and P. Labine. NACE (Houston), 1993.
4. Recommendation for repair strategies for concrete structures damaged by reinforcement corrosion, RILEM TC 124, *Mater. Struct.*, 1994, **27**, 415.
5. Erhaltung von Betonbauwerken, Richtline SIA 162/5, Schweiz. Ingenieur und Architektenverein Zürich 1996.
6. B. Elsener, D. Flückiger, H. Wojtas and H. Böhni, Methoden zur Erfassung der Korrosion von Stahl in Beton. VSS Forschungsbericht Nr. 521 (1996), Eidg. Verkehrs- und Energiewirtschaftsdepartement — Bundesamt für Strassenbau. (Methods to detect corrosion of rebars in concrete EVED Swiss Federal Highway Agency Research Report No. 521 (1996) Publ. Verein Schweizer Strassenfachleute (VSS) Zurich, Switzerland (in German)).
7. D. F. Griffin, Corrosion inhibitors for reinforced concrete, in *Corrosion of Metals in Concrete*, ACI SP-49. American Concrete Institute, 1975, p. 95.
8. N. S. Berke, Corrosion inhibitors in concrete, *Corrosion '89*, Paper 445, NACE, Houston, Tx, 1989.
9. J. M. Gaidis and A. M. Rosenberg, *Ceme. Concr. Aggreg.*, 1987, **9**, 30.
10. B. El-Jazairi and N. Berke, *Corrosion of Reinforcement in Concrete Construction*, Ed. C. L. Page. Elsevier Applied Science, London, 1990, p. 571.
11. Concrete Bridge Protection and Rehabilitation: Corrosion Inhibitors and Polymers, SHRP Report S-666, National Research Council, Washington DC, 1993.

12. B. D. Prowell, R. E. Weyers and I. L. Al-qadi, Evaluation of corrosion inhibitors for the rehabilitation of RC structures, in *Concrete 2000*, Eds R. K. Dhir and M. R. Jones. E&FN Spon, 1993, p. 1223.

13. N. S. Berke and T. G. Weil, World wide review of corrosion inhibitors in concrete, in *Advances in Concrete Technology*, Ed. V. M. Malhotra. Ottawa, Canada CANMET 1992, p. 899.

14. M. G. Arber and H. E. Vivian, *Australian J. Appl. Sci.*, 1961, **12**, 339.

15. K. W. Treadaway and A. D. Russel, *Highways and Public Works*, 1968, **36**, 40.

16. M. Hynes and B. Malric, Use of migratory corrosion inhibitors, *Construction Repair*, July / August 1997, p. 10.

17. C Alonso, C. Andrade, C. Argiz and B. Malric, *Cem. Concr. Res.*, 1992, **22**, 869.

18. C. Andrade, C. Alonso, M. Acha and B. Malric, *Cem. Concr. Res.*, 1996, **26**, 405.

19. P. Schmalz and B. Malric, Korrosionsbekämpfung im Stahlbeton durch Inhibitoren auf MFP Basis, Erhaltung von Brücken, SIA Dokumentation D 099 (1993) p. 65, Schweiz. Ingenieur und Architektenverein, Zürich. (Corrosion Mitigation of Reinforced Concrete Structures by MFP Based Inhibitors in "Erhaltung von Brücken" SIA Dokumentation D099 (1993), p. 65. Publ. Schweiz. Ingenieur und Architektenverein, Zurich, Switzerland (in German).)

20. P. Gassner, B. Malric, F. Hunkeler and B. Elsener, Korrosionsbekämpfung im Stahlbeton durch Inhibitoren auf MFP Basis, Draft of Final Report (1998). (Corrosion Mitigation of Reinforced Concrete Structures by MFP Based Inhibitors, EVED, Swiss Highway Agency Research Report (1998) Publ. Verein Schweizer Strassenfachleute (VSS), Zurich, Switzerland (in German).)

21. P. Annen and B. Malric, Surface applied inhibitor in rehabilitation of the Peney Bridge, Geneva (CH), in *Bridge Management 3*, Eds E. Harding, G. A. R. Parke and M. J. Ryall. E&FN Spon, London (1996), 437.

22. U. Mäder, A new class of corrosion inhibitors, in *Corrosion and Corrosion Protection of Steel in Concrete*, Ed. N. Swamy, 1994, **II**, 851.

23. European Patent Application No. 8630438.2, Publication No. 0 209 978, published 28.01.87 Bulletin 87 / 5.

24. European Patent Specification No. 87903356.1, Publication No. 0 305 393 B1 published 15.05.87. International Patent Application PCT / GB87 / 00339.

25. G. E. Fodor, The inhibition of vapour phase corrosion: a review. *Reviews on Corrosion Inhibitor Science and Technology*, Eds A. Raman and P. Labine. NACE, Houston (1993), page II-17.

26. B. A. Miksik, Use of vapour phase corrosion inhibitors for corrosion protection of metal products, Reviews on Corrosion Inhibitor Science and Technology, Eds A. Raman and P. Labine. NACE Houston (1993), 11–16.

27. A. Phanasgaonkar, B. Cherry and M. Forsyth, Corrosion inhibition properties of organic amines in a simulated concrete environment, in *Proc. Int. Conf. on Understanding Corrosion Mechanisms of Metals in Concrete — A Key to Improving Infrastructure Durability*. Massachusetts Institute of Technology, MIT (Cambridge, USA) 1997, section 6.

28. P. H. Laamanen and K. Byfors, Corrosion inhibitors in concrete — Alkanolamine based inhibitors — state of the art report, Nordic Concrete Research No. 19, 2 / 1996, Norsk Betongforening, Oslo (1996).

29. B. A. Miksik, M. Tarvin and G. R. Sparrow, Surface analytical techniques in evaluation of VCI organic corrosion inhibitors on the surface chemistry of metals, *Corrosion '89*, Paper 607, NACE, Houston, Tx, 1989.

30. C. R. Brundle, D. Grunze, U. Mäder and N. Blank, *Surf. Interface Anal.*, 1996, **24**, 549.

31. G. S. Bobrowski, M. A. Bury, S. A. Farrington and C. K Nmai, Admixtures for inhibiting corrosion of steel in concrete, US, Patent No. 5.262.089, 16.11.1993.

32. C. K. Nmai, S. A. Farrington and G. S. Bobrowski, *Concr. Int.*, American Concrete Institute 1992, **14**, 45.

33. M. Büchler, B. Elsener and H. Böhni, A migrating corrosion inhibitor blend for reinforced concrete. Part I: prevention of corrosion. Submitted to CORROSION/NACE.

34. A. Rossi, B. Elsener, M. Textor and N. D. Spencer, Combined XPS and ToF-SIMS analyses in the study of inhibitor function — organic films on iron, *Analysis*, 1997, **25**, 5 M30.

35. B. Elsener, M. Molina and H. Böhni, *Corros. Sci.*, 1993, **35**, 1563.

36. B. Elsener and H. Böhni, Half cell potential measurements — from theory to condition assessment of RC structures, in *Proc. Int. Conference on Understanding Corrosion Mechanisms of Metals in Concrete — A Key to Improving Infrastructure Durability*. Massachusetts Institute of Technology, MIT (Cambridge, USA) 1997, paper No. 3.

37. B. Elsener, L. Zimmermann, D. Bürchler and H. Böhni, Repair of reinforced concrete structures by electrochemical techniques — field experience, in *Proc. EUROCORR '97*, Trondheim, Norway 1997, **1**, 517–522.

38. J. Broomfield, The pros and cons of corrosion inhibitors, in *Constr. Rep.*, July/August 1997, 16.

39. U. Nürnberger and W. Beul, *Werkst. Korros.*, 1991, **42**, 537.

Part 3

Corrosion Rate

Measurement

7

Performance Testing of Corrosion Inhibitors for Concrete using Mortar Samples

J. VOGELSANG, U. ESCHMANN and G. MEYER

Sika Chemie GmbH, Kornwestheimer Str 107, D- 70439 Stuttgart, Germany

ABSTRACT

Several preparation techniques for the preparation of mortar samples for corrosion rate measurements of steel in chloride contaminated concrete are presented and the accompanying errors described. It was not possible to achieve a mortar sample preparation with a steel electrode including electrical connection embedded completely in the mortar. The most significant errors came from crevice corrosion under the sealants and coatings applied for limiting the exposed steel surface and for electrical insulation. The only successful preparation technique was that described by Andrade in 1980 in which an insulation tape was wrapped around the steel rods at the exposed ends pushed into the mortar, the steel outside the mortar being covered with Vaseline to prevent atmospheric corrosion.

1. Introduction

Chloride induced corrosion of usually passive reinforcing steel is a well known problem, especially where de-icing salts are used or chloride contaminated aggregates are incorporated into the concrete [1].

Corrosion inhibitors for concrete are becoming increasingly important for reducing corrosion damage brought about by chlorides or carbonation. A first approach for screening of possibly inhibitive substances is a simple immersion test [2]. Electrochemical inhibitor testing in solution is the next step in finding effective but environmentally friendly inhibitors.

Obviously, companies have to check the efficiency of their products in 'natural' surroundings whether they are producing corrosion inhibitors to be applied as admixtures in the mixing water during concrete production or corrosion inhibitors to be applied as impregnation for structures when these first start to show damage. In both cases efficiency testing is required in mortars/concrete since corrosion tests using methods much closer to practice than testing in solution are necessary.

In many publications the description of the sample preparation is given. Sometimes these samples are called 'lollipops', and are prepared by — more or less — completely embedding an iron wire or rod, as the working electrode, into the mortar. A basic design is shown in Fig. 1(a). Four other possibilities of mounting the steel bar have been published and can be seen in Fig. 1(b)–(e) [3–7]. Another sample preparation technique described by Elsener et al. [8] seems to be promising, with good results obtained by these authors, but the connection tools are not available commercially.

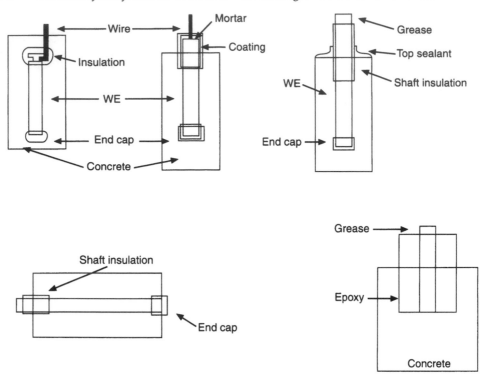

Fig. 1 *The various types of electrode preparation.*

Therefore some requirements should be discussed for preventing unintended side effects such as contact corrosion or crevice corrosion. If the working electrode is to be embedded completely in the mortar, it will be necessary to have a wire for electrical contact for the measurement of the potential and the current. However, with this, contact corrosion can occur if the sealing of the electrical contact between the steel working electrode and the copper wire is not sufficiently watertight and an ion flux between the steel and the copper can — according to the electromotive series — electrochemically promote the dissolution process of the steel. It is also important that the exposed area of working electrode is known and that this homogeneous area should be defined using, as far as possible, inert materials for electrical insulation and corrosion protection of the unexposed area. Only by knowing the value of the exposed area is it possible to calculate the absolute corrosion current density I_{corr} for example as μAcm^{-2} (the sense in giving a corrosion rate in the case of pitting corrosion — the main corrosion process in an alkaline media — will not be discussed here).

One of the most common and undesired effects is crevice corrosion under the sealant at the edges and under the insulation of the electrical connection to the working electrode (Fig. 2). The measurement of the corrosion rate is thus highly affected by crevice and contact corrosion.

In this contribution we discuss sample preparations that can give the most erroneous results and our successful approach to overcoming these. Using this successful sample preparation we investigated some inhibitive admixtures. The results of these tests are also presented here.

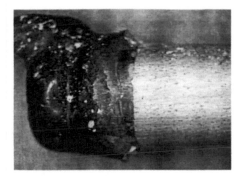 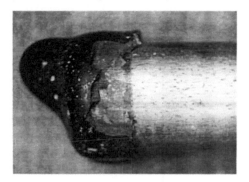

Fig. 2 *End cap and insulation of the electrical connection, both with crevice corrosion; the coating was easy to remove and corrosion products were found. Method 1a.*

2. Experimental

Tables 1(a), 1(b) and 1(c) describe the various insulation materials corresponding to the drawings in Figs 1(a), 1(b) and 1(c) respectively (electrode designs corresponding to Figs 1(d) and 1(e) were not investigated in this work). Thus, Table 1(a) lists five types of insulation and end cap used for design (a) in Fig. 1. Table 1(b) lists the shaft materials insulation, end cap and top sealant design used for design (b) in Fig. 1. Table (c) lists the mortar type, coating and end cap for design (c). In design type (b) and (c) the height of the lollipop was 6 cm, in design (a) 10 cm and the diameter was 3.5 cm for each design. The dimensions of the exposed steel electrode were 3 cm high with a diameter of 1 cm.

The steel was degreased with xylene and was free of rust. The mortar was prepared with a sand type as defined in DIN/EN 196/1 and Portland cement type PZ 35 in a ratio of 3:1 with a water cement ratio of 0.5. The sodium chloride was added to provide a concentration of 3% of the cement weight for all preparations. At least four individual samples were prepared for each preparation material to avoid statistical errors.

(For the evaluation of the inhibitors the chloride content and the inhibitor concentration are given in the discussion of the results or in the figure captions of Figs 6 and 7.)

After two days in the humidity chamber the samples were removed from their formwork and were cured for another two weeks in the humidity chamber at 23°C/ 100% R.H. Before measuring the I_{corr} the samples were kept at 23°C/80% R.H. for a further two weeks.

After one hour of wetting in a 0.01M KNO$_3$ solution, electrochemical impedance spectra (EIS) were taken in the frequency range of 1 Hz to 100 kHz with an amplitude of 5 mV to check whether conductivity of the mortar was sufficient. The I_{corr} was determined with a Tafel type measurement from −300 mV vs ocp (open circuit potential) to +200 mV vs ocp by measuring the current after a maximum waiting time of 3 min. every 5 mV to achieve constancy/steady state in the current/ electrochemical system. After the measurement the samples were broken and the sealants and insulations examined visually.

Table 1(a)

Number	Insulation	End cap
1	Two pack, solvent free high build polyurethane – coating	Like insulation
2	One component moisture curing polyurethane clearcoat, two pack solvent free high build polyurethane	Like insulation
3	Sand blasting of areas to be coated Coating like No. 2	Like insulation
4	Two pack epoxy coating containing mineral fillers and anthracene oil, product specially developed for application on steel structures in water ways. Substituting coal tar epoxy.	Like insulation
5	Electroplaters tape (3M) according ASTM G 109-92, additionally with adhesive sealant, silicon rubber, acetate cured	Like insulation

Table 1(b)

Number	Shaft Insulation	End Cap	Top sealant
1	Epoxy clear coat	Like shaft	Two pack epoxy coating for concrete
2	Plastic tape Scotch Super 33+ (3M)	No end cap	Two pack epoxy coating for concrete

Table 1(c)

Number	Mortar-type	Coating (for insulation)	End cap
1	Cementitious epoxy-system with low VOC-content	(a) Two pack epoxy coating for steel and concrete (b) Two pack epoxy tooling resin	Like insulation
2	Polymer modified one component mortar containing silica fume	(a) Two pack epoxy coating for steel and concrete (b) Two pack epoxy tooling resin	Like insulation
3	Three pack cement based epoxy modified mortar	(a) Two pack epoxy coating for steel and concrete (b) Two pack epoxy tooling resin	Like insulation
4	Standard cement with siliceous fine aggregates, mixing water with styrene acrylic dispersion	(a) Two pack epoxy coating for steel and concrete (b) Two pack epoxy tooling resin	Like insulation

3. Results and Discussion

The observations for the individual preparation technique as described in Table 1 are listed in Table 2. It can be seen that all of the preparation techniques gave rise to crevice corrosion, except for method 2 from Table 1 (b).

In Fig. 2 two pictures are presented showing the corrosion products under the insulation and the end cap of method 1(a). The coatings very often showed delamination in such a way that the end cap could easily be removed. Corrosion products were found under the coating, the area sometimes was wet and contact corrosion occurred near the copper wire. It was not possible, even by sand blasting, to obtain sufficient adhesion of the coatings. The use of a very thick epoxy cover, which altered the working area from the cylinder area to the base area as shown in Fig. 1(e), was also not successful. Despite applying an epoxy material which contracted during the curing process, a crevice was formed and was possibly initiated when grinding the working area before it was brought into contact with the fresh mortar.

The use of a sealant in addition to the tapes did not give any improvement — contact corrosion and ordinary corrosion was still present. Only the amount of crevice corrosion was reduced significantly. So our desired electrode geometry had to be abandoned and the electrode design of Fig. 1(b) was further considered [4]. In this, the end cap was still a problem and our final successful preparation did not have

Table 2

Number		Corrosion on working electrode surface	Crevice corrosion	Contact corrosion
1a	1	No	Yes	Yes
	2	No	Yes	Yes
	3	Yes	Yes	Yes
	4	No	Yes	Yes
	5	Yes	Yes	Yes
1b	1	Yes	Yes	No
	2	Yes	No	No
1c	1	Yes	Yes	Yes
	2	Yes	Yes	Yes
	3	Yes	Yes	Yes
	4	Yes	Yes	Yes

this cap. The best way of sample preparation was found to be method 2 of Table 1(b) with the plastic tape (Scotch Super 33+ from 3M) and without end cap.

Sometimes, the presence of crevice corrosion (anodic process) protected the working area by some cathodic polarisation, so that no corrosion traces were found on the working area, even when the chloride content should have been high enough to cause pitting corrosion.

However, adding chloride to the mixing water is not very close to practice because it is only with chloride contaminated aggregates that the chloride is present from the beginning of the curing. Normally, the inhibitor is present first and the chloride has to migrate from outside to the rebar. The inhibitor is then able to develop its protective action first, i.e. before chloride ions attack the passive steel. It could therefore be advantageous to prepare the mortar without chloride addition and then have the chloride attack through the mortar by diffusion. But this way could also show some disadvantages. Firstly, the diffusion will not be uniform over the whole sample because of micro cracks or different pore sizes. Secondly, the samples will show differences in diffusion rate when compared to each other and so it would not be easy to distinguish between a real inhibitive effect and only a delayed or reduced diffusion rate of the chloride. Further, these experiments will need too much time because the diffusion rate of chloride in concrete or mortars is very low. From this it can be seen that the chloride addition to the mixing water is a good compromise between practical and temporal demands.

Two series with different inhibitors and chloride contents were prepared with method 2 of Table 1(b). Series A was prepared with five groups of four lollies, three with different inhibitors and 3% sodium chloride, one with 3% sodium chloride only and one group without inhibitor and without chloride. Series B was made with only 1.5% sodium chloride, which is a concentration much closer to practical cases. The corrosion current densities were obtained as usual for Tafel-type measurements. In Fig. 3, three typical curves are shown, one sample with and one without chloride, the third curve demonstrates the efficiency of the inhibitor under the condition of chloride contamination. The curves are significantly different by nearly two orders of magnitude. Normally, the individual values of identically prepared samples showed considerable fluctuations due to the stochastic nature of pitting corrosion. Nevertheless, using at least four samples in each case it was possible to distinguish between effective and non-effective inhibitors and the repeatability is fair. The assumption that I_{corr} varies with the electrolyte resistance of the mortar should be questioned in view of Fig. 4 and the measured I_{corr} values given for the respective samples.

The low frequency impedance allows one to detect corrosion significantly, but the measurement is very time consuming, so the easier and faster measurable Tafel plots were preferred in this work. The good correlation of Tafel plots and Bode plots can be seen in Fig. 3 and Fig. 5. The sample with chloride showed the lower polarisation resistance (Fig. 5) and accordingly I_{corr} was also higher. The precise values are not discussed here because the determination of corrosion rates using EIS under the condition of localised and non uniform corrosion is doubtful. For our investigation it was sufficient to see the qualitative agreement of the two methods.

In Fig. 6 the results of series A are shown. With an amount of 3% sodium chloride one inhibitor was not effective and only a very high amount of sodium nitrite was

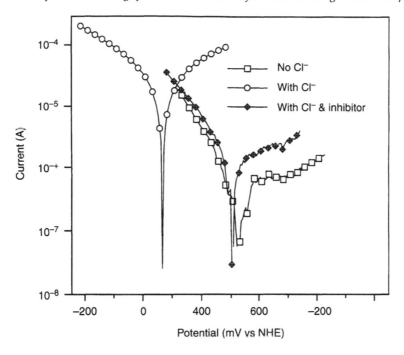

Fig. 3 *Tafel plot of three lollipops with and without 3% NaCl and with chloride and 3% inhibitor.*

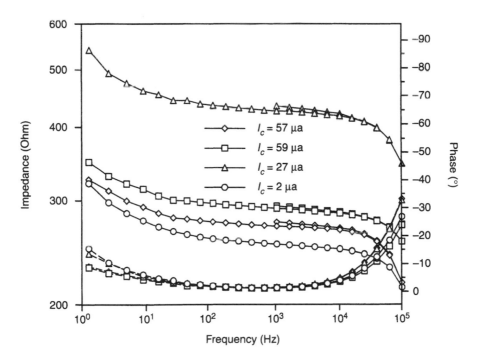

Fig. 4 *Bode plot of impedance spectra obtained from four samples containing 3% NaCl; the correlation of corrosion current density with mortar conductivity seems to be questionable.*

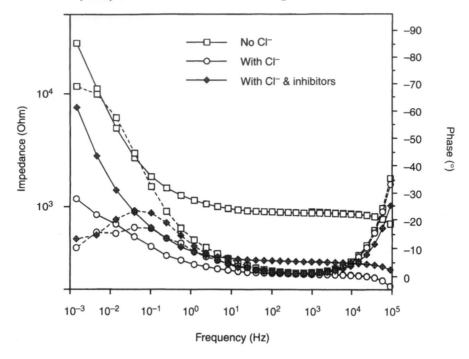

Fig. 5 *Significant differences also in Bode plots between samples with and without 3% NaCl; measuring time is too short for the determination of the polarisation resistance; third curve is with 3% inhibitor and NaCl.*

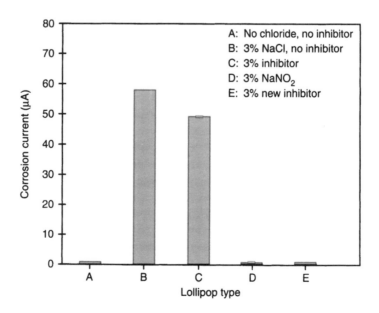

Fig. 6 *Three different inhibitors tested in mortar containing 3% NaCl. Lollipop type A = without chloride.*

able to prevent corrosion. A new inhibitor reduces corrosion significantly. But it can not be expected, that an inhibitor will show the same efficiency as at passivator, like nitrite, if both are applied at the same amount. The inhibitors were 30% solutions of the inhibitive substance, so this ranking could be turned round if the same equivalent of inhibitor and passivator were used. In Fig. 7 we wanted to see at a more realistic chloride level whether two inhibitors work and which is the inhibitor concentration to be used. It turned out that 3% of inhibitor is sufficient for corrosion protection, but the non-realistic concentration of 5% is too much because the inhibitor starts to affect the mortar properties.

4. Conclusions

To develop effective inhibitors it is very important to find a technique for the preparation of a mortar test sample which allows one to distinguish between corrosion and no corrosion on a possibly large scale. We found that it is quite impossible to create samples with completely embedded working electrodes and electrical connections inside the mortar. In our opinion it is impossible to use epoxy coatings or resins as effective sealants since we always found crevice corrosion. The same can be stated for polyurethanes and all of the coating materials used here. It must be pointed out that all the coating materials used are performing extremely well as corrosion protective coatings for rural, urban and industrial atmospheres and some are used offshore. So the conditions in such a fresh mortar with the very high chloride content is a crucial environment for an organic coating when this is only partly applied to the steel surface.

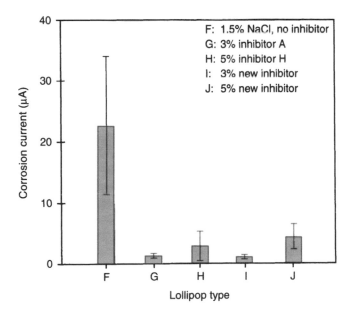

Fig. 7 *Lollipop formulations with 1.5% NaCl; two inhibitors with different amounts (3 or 5%) of inhibitor added to the mixing mortar.*

Another point of importance is the galvanic corrosion arising from the copper wire–steel electrode couple because it is impossible to seal the connection effectively over such a long conditioning period (e.g. 4 weeks) with the techniques investigated in this work.

Using a method in which a steel electrode was insulated by a plastics tape with no end cap and a two pack epoxy coating for the top sealant, it was possible to prepare suitable mortar samples which provided reliable results that were well understood and free from crevice or contact corrosion effects.

References

1. U. Nürnberger, Korrosion und Korrosionsschutz im Bauwesen, Band 1, Bauverlag Wiesbaden, Berlin, 1995.
2. J. Vogelsang and G. Meyer, Electrochemical properties of concrete admixtures, in *Corrosion of Reinforcement in Concrete Construction*, Eds C. L. Page, P. B. Bamforth and J. W. Figg. The Royal Society of Chemistry, special publication No. 183, 1996.
3. P. Lambert, Ph.D. Thesis, 1983. Aston University, Birmingham, UK.
4. J. A. González, S. Algaba and C. Andrade, *Brit. Corros. J.*, 1980 **15** (3),135–139.
5. F. Wenger and J. Galland, *Mater. Sci. Forum*, 1989, **44 & 45** 375.
6. P. Toumey and N. Berke, *Concr. Int.*, 1993, Apr. 57.
7. P. Lambert, C. L. Page and P. R. W. Vassie, *Mater. Struct.* 1991, **24**, 351.
8. B. Elsener and H. Böhni, *Mater. Sci. Forum*, 1986, **8**, 363.

Dependence of Corrosion Rate of Rebars on Climatic Parameters in Outdoor Concrete Structures Contaminated with Chlorides

C. ANDRADE, C. ALONSO and J. SARRIA

Institute of Construction Sciences "Eduardo Torroja", CSIC, E-28033 Madrid, Spain

ABSTRACT

The main concrete parameter influencing the corrosion rate of rebars when they are depassivated is the moisture content of the concrete. In concrete structures exposed to outdoor atmospheres, this depends on the climatic cyclic variations of relative humidity and temperature which continuously change during the day and night and with seasonal variations. The corrosion rate also changes but the value recorded at a precise moment may not correspond to the actual hygrometric situation. This makes the implementation of corrosion rate values measured on-site into predictive models difficult, and calls for more detailed studies and statistical treatments. In this paper changes in the corrosion rate in concrete structures submitted to outdoor exposure and not sheltered from rain are monitored. More detailed measurements were made during certain periods. Simultaneous recording of temperature, relative humidity, resistivity, corrosion potential and corrosion rate enabled changes in the complexity of interacting processes to be followed. Some specific trends are identified from which it can be deduced that the seasonal changes in temperature control the main parameters involved, although a general relation between temperature and corrosion rate could not be found.

1. Introduction

Once a rebar depassivates as a result of the action of chlorides or the pH drop induced by carbonation, the corrosion process starts to develop. The further development of the corrosion rate was studied some time ago [1] by means of the Polarisation Resistance technique [2]. It is well established that the moisture content of the concrete is the main parameter controlling the rate of the process. Thus, when the concrete is dry, the corrosion drops to negligible values (below $0.1 \ \mu Acm^{-2}$) and increases when the humidity increases [3].

However, very few data have been published on the influence of the climatic changes, and in particular of the temperature, $T°$, on the corrosion rate [4,5]. All the studies that have been reported have been made in controlled conditions of R.H. and $T°$ in the laboratory. Only recently, have data been published [6] on the effect of the natural climatic day–night cycles on the corrosion rate V_{corr} of a concrete structure contaminated with chlorides. This work found that the variations of V_{corr} seemed not to follow any specific trend. This behaviour was attributed to the fact that in natural conditions, the R.H. and $T°$ are in continuous change inside the concrete and equilibrium is not achieved.

In the present paper the study is continued and additional results are presented which could help in the understanding of the complex multiple effect that simple continuous $T°$ cycles induce in the corrosion process.

2. Experimental

Two kinds of structures were selected. One had been artificially contaminated with chlorides from the concrete mix and the other was a structure located in a garden and suffering carbonation.

The concrete structure containing chlorides is the beam shown in Fig. 1 [6]. It had been fabricated with 360 kg of OPC (Ordinary Portland Cement) per cubic meter of concrete and a w/c (water/cement) ratio of 0.7; 3% of $CaCl_2$ per weight of cement was added to the mix to promote corrosion. The beam was fabricated outdoors and water cured during three days. It was then exposed to the action of the climate.

The other is an artistic structure (Fig. 2) about 35 years old whose rebars are covered by rust attributed to the carbonation of the cover; despite this no cracking or spalling has been produced. This structure is wetted at least three times per week when the garden is watered.

The electrochemical parameters, corrosion potential, E_{corr}, resistivity and corrosion rate, I_{corr}, were measured by means of a portable corrosion rate meter, Gecor 06, having sensorised confinement of the current [7]. In addition, R.H. measurements were continuously monitored by means of different R.H. sensors [6] placed in some cavities that had been made in the beam and in the artistic structure.

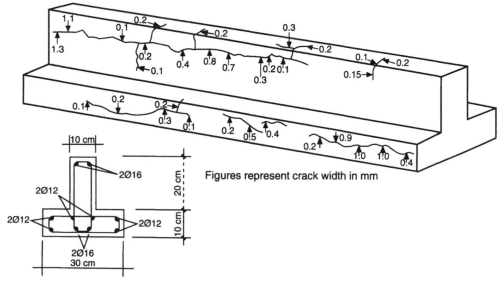

Fig. 1 *Appearance of the beam used for the tests. Crack widths are indicated, as well as the size of the beam.*

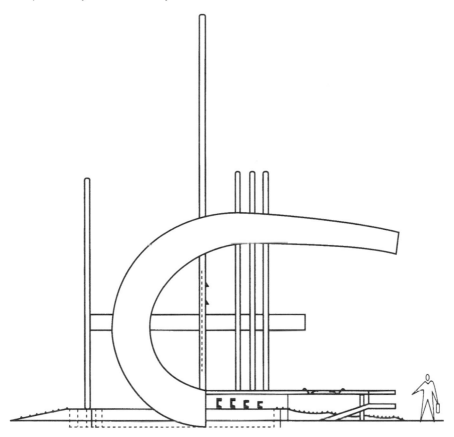

Fig. 2 *Transverse view of the artistic structure.*

3. Results
3.1. Chloride-containing Structure

Figure 3 shows the record of the $T°$ variations during two years in the external (out) atmosphere and inside the cavity (in) of the beam. It can be seen that in and out values of $T°$ are very similar.

However, this is not the case with the R.H. (Fig. 4). As has been previously reported [8], while the external day-night R.H. changes are significant, inside the beam the R.H. remains quite stable. This trend is general and only seasonal variations induce a definite R.H. change, with the R.H. being lower in spring–summer and higher in autumn–winter.

The combined effect of $T°$ and R.H. can be better visualised in Fig. 5, where the R.H. inside the beam cavity and outside have been plotted in a psicrometric abacus. Each point represents the mean R.H. during each month over two years. It can be seen that in the external atmosphere the changes in temperature induce significant changes of R.H. since the total humidity remains unchanged. However, inside the cavity, the temperature changes induce phenomena of condensation-evaporation which enable the R.H. to remain quite constant.

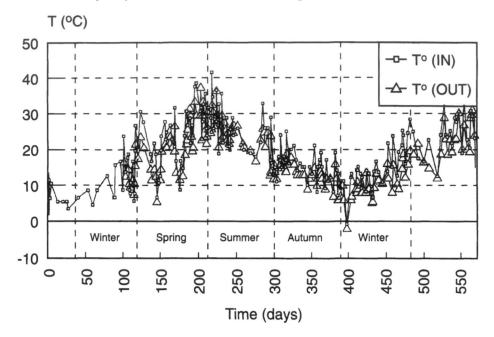

Fig. 3 *Changes of temperature with time in the chloride contaminated structure.*

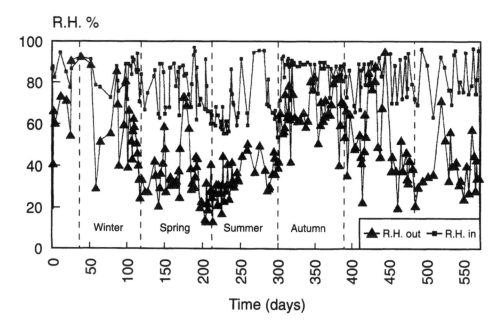

Fig. 4 *Internal (cavity) (■) and external (▲) RH in the beam.*

Figure 6 indicates that no precise trend can be identified in I_{corr} during the two years testing. It appears that R.H. and $T°$ do not have any direct relation to the

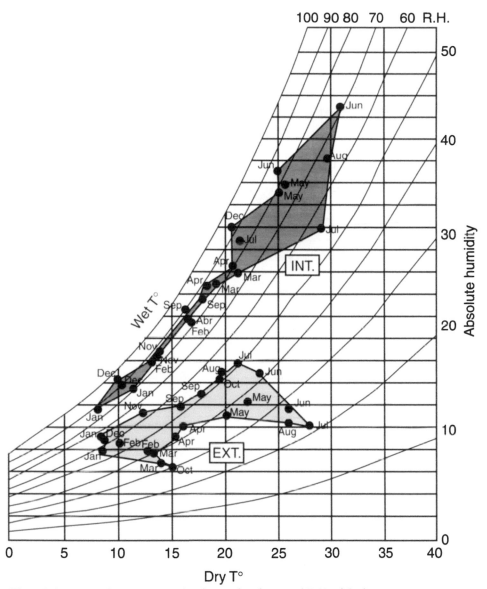

Fig. 5 *Psicometric abacus representing internal and external R.H. of the beam.*

corrosion rate. This unexpected fact warranted the need for further studies to ascertain the I_{corr} behaviour.

Thus, Fig. 7 depicts the changes in I_{corr} during a single day from the morning to the evening. It is clear here that the $T°$ has an effect on the I_{corr} from which it can be concluded that in the short term daily cycle, the $T°$ does influence the I_{corr}.

In a similar manner the effect of rain periods has been studied. Thus, in Fig. 8 an example is shown of how after raining (in day 5 in the figure), a high I_{corr} value can be achieved (in day 10 in the figure). Therefore, a higher humidity also induces a higher corrosion rate.

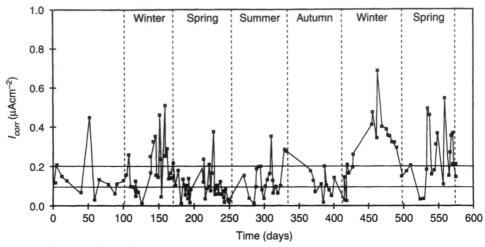

Fig. 6 *Record of* I_{corr} *value over two years.*

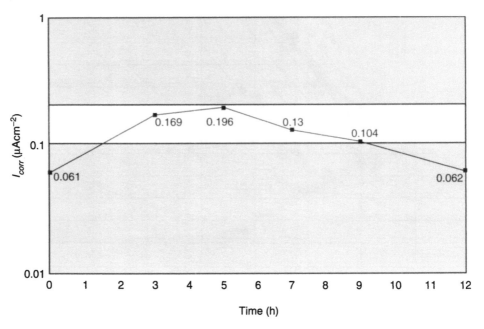

Fig. 7 *Record of* I_{corr} *values for a duration of 1 day.*

3.2. Carbonated Structure

In the case of the old carbonated structure, Fig. 9 shows the changes in its out and in-R.H., which is very high as is expected from its frequent wetting. However despite this, the I_{corr} values remain very low as shown in Fig. 10. This behaviour is attributed to the dense layers of rust which could be seen when making a small hole for direct observation of the rebars. It seems that the old rust and the high humidity content resulted in a negligible value of the I_{corr}.

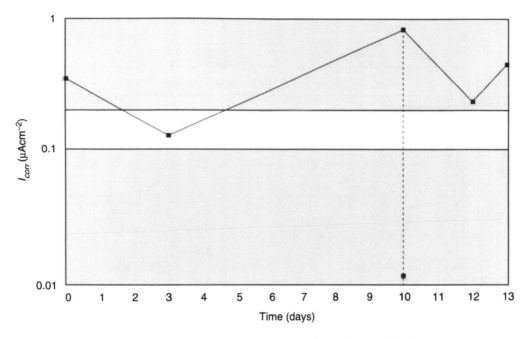

Fig. 8 I_{corr} *values pointing out a maximum (in day 10) after a rain period in day 5.*

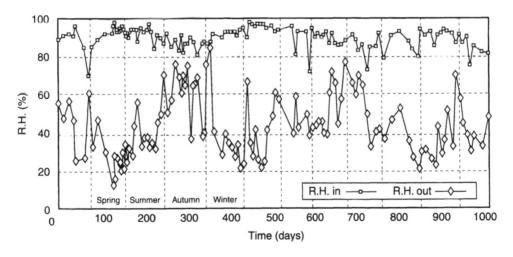

Fig. 9 *Averaged* I_{corr} *values as a function of R.H. and $T°$ ranges.*

4. Discussion

The dependence of I_{corr} from climatic parameters seems very complex and it is not possible at present to offer a mathematical expression relating I_{corr} to R.H. and $T°$.

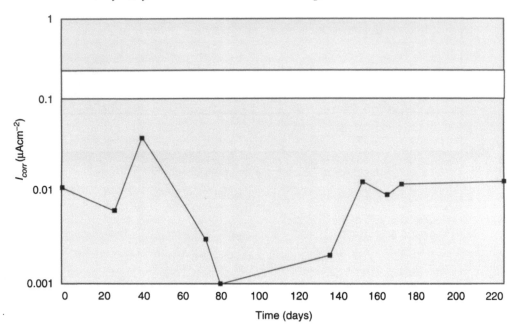

Fig. 10 *Changes in I_{corr} in the carbonated structure.*

On the one hand, $T°$ changes induce R.H. variations which follow different trends from those in the atmosphere, and on the other hand, $T°$ has several effects which may counterbalance each other in respect of I_{corr}. Thus, if $T°$ increases for instance, the expected influences are [9]:

(a) a decrease in the oxygen content of the pore solution,

(b) a decrease in the pore solution resistivity,

(c) an increase in the Cl^-/OH^- ratio, and

(d) a certain evaporation (increase in the pore R.H. but also increase in the concrete resistivity).

All these changes point to the fact that so far the trend of the corrosion potential and corrosion rate are not predictable.

The best fit is still found between I_{corr} and concrete resistivity [10]. Between both parameters the expression I_{corr} $(\mu Acm^{-2}) = 10^4 \rho(\Omega cm^{-1})$ is in general fulfilled. However, the relation between Icorr and resistivity, ρ, when generalised for numerous structures, presents a high scatter indicating that other parameters must influence the relation.

5. Conclusions

It can be concluded that further studies have to be carried out to clarify the precise role of R.H. and $T°$ in the corrosion rate. From present results the main conclusions that can be drawn are:

1. As has been previously observed, the R.H. inside the concrete remains quite stable in spite of the dramatic day–night changes of external R.H. and $T°$. However, the seasonal changes of $T°$ during the year, as well as the periods of rain, do significantly influence the R.H. inside a cavity.

2. The electrochemical corrosion parameters, E_{corr} and I_{corr} do follow the day-night changes. However, due to the opposing effects of $T°$ variations on the concrete and in the pore solution physico-chemical characteristics, a general relation of these with the $T°$ could not so far be found. One reason for this is that the periods of rain vary the total moisture content of the concrete.

References

1. C. Andrade and J. A. González, Quantitative measurements of corrosion rate of reinforcing steels embedded in concrete using polarization resistance measurements,*Werkst. Korros.*, 1978, **29**, 515–519.

2. J.A. González and C. Andrade, Effect of carbonation, chlorides and relative humidity on the corrosion of galvanized rebars embedded in concrete, *Brit. Corros. J.*, 1982, **17**, (1) 21–28.

3. G. K. Glass, C. L. Page and N. R. Short, Factors affecting the corrosion rate of steel in carbonated mortars, *Corros. Sci.*, 1991, **32**, 1283–1294.

4. W. López, J. A. González and C. Andrade, Influence of temperature on the service life of rebars, *Cem. Concr. Res.*, 1993, **23**, 1130–1140.

5. P. Schiessl and M. Raupach, Influence of temperature on the corrosion rate of steel in concrete containing chlorides, 1*st Int. Conference of Reinforced Concrete Materials in Hot Climates*. United Arab Emirates University, Al Ain U.A.E., Al Ain, 1994, 537–549.

6. C. Andrade, J. Sarría and C. Alonso, Statistical study on simultaneous monitoring of rebar corrosion rate and internal relative humidity in concrete structures exposed to the atmosphere, *4th Int. Symp. on Corrosion of Reinforcement in Concrete Construction*, Cambridge, UK, 1996. Soc. Chem. Ind., London, 1996, p. 233–242.

7. S. Felíu, J. A. González, S. Felíu Jr. and C. Andrade, Confinement of the electrical signal for in-situ measurement of polarization resistance in reinforced concrete, *Mater. J. ACI*, 1990, **97**, (5), 457–460.

8. E. J. Sellevold, Resistivity and humidity measurements of repaired and non repaired areas in Gimsøystraumen bridge, *Int. Conf. on Repair of Concrete Structures*, Svolvaer (Norway) May 1997, Norwegian Road Research Lab. P.O. Box 8142 Dep.N-0033, Oslo.

9. C. Andrade, C. Alonso and J. Sarría, Influence of relative humidity and temperature on the on-site corrosion rates, *4th CANMET/ACI Int. Conf. on Durability of Concrete*, Sydney (Australia), August 1997, Publ. by ACI. SP-170.

10. C. Alonso, C. Andrade and J. A. González, Relation between concrete resistivity and corrosion rate of the reinforcements in carbonated mortar made with several cement types, *Cem. Concr. Res.*, 1988, **18**, 687–698.

9

Corrosion Rate of Steel in Concrete — From Laboratory to Reinforced Concrete Structures

B. ELSENER

Institute of Materials Chemistry and Corrosion, Swiss Federal Institute of Technology, ETH Hönggerberg, CH-8093 Zürich, Switzerland

ABSTRACT

Condition assessment, control of the efficiency and durability of restoration work and service life prediction of reinforced concrete structures need rapid, non-destructive techniques to assess corrosion of the rebars and provide a quantitative measure of the instantaneous corrosion rate. Electrochemical techniques to measure corrosion rates of steel in concrete are based on the determination of the polarisation resistance, R_p, and in laboratory experiments comparable results are obtained with different techniques. Polarisation resistance measurements on-site have to take into account the following points: non-homogeneous current distribution between counter electrode and rebar network, changes of R_p with daily and seasonal fluctuations in temperature and humidity, influence of the type of instrument and measurement principle. Finally, the conversion of the experimentally measured $R_{p,eff}$ to a specific polarisation resistance $R_p{}^*$ (or the instantaneous corrosion rate) requires a correct compensation of the ohmic resistance and a knowledge of the actively corroding area.

1. Introduction

Corrosion of the rebars due to chloride attack and / or carbonation is the main cause of damage and early failure of reinforced concrete (RC) structures with enormous costs for maintenance, restoration and replacement worldwide. Rational planning of the restoration, control of the efficiency of repair work and service life prediction of RC structures need rapid, non-destructive on-site techniques that detect corrosion of the rebars at an early stage, define adequately which areas of structures are corroding and provide a measure of the corrosion rate.

Whereas the location of corroding areas can be achieved by half-cell potential measurements, the correct interpretation of the measured data and determination of the 'corrosion rate' on RC structures is not straightforward.

1.1. Location of Corroding Areas — Half-cell Potential Mapping

The use of electrochemical potentials to determine areas of corrosion risk of reinforcing steel in concrete was pioneered in the United States [1,2] and resulted in the development of an ASTM standard (ASTM C876-91). Today potential mapping is 'state of the art' to locate corroding zones precisely ([3–5] and references cited therein). The extent of any corrosion problem of the structure being investigated can be mapped prior to more detailed and costly examination and repair. Field experience on a large number of bridge decks and substructures has shown [4,5] that an absolute criterion to indicate corrosion (e.g. –350 mV SCE) cannot be applied: changes in concrete resistivity influence potential readings on the surface [4] and also different pH values of the concrete shift the potential of passive steel [6]. Even potential readings taken on the concrete surface can be misinterpreted, for example, lack of oxygen in very wet, dense or polymer-modified concrete leads to negative potentials without corrosion of the rebars [7]. It is obvious that no universal relation between half-cell potential at the concrete surface and corrosion rate of the embedded rebars can be established, but for a single structure the half-cell potentials have shown to correlate well with the actual state (loss in cross section) of the rebars [5]. The corrosion rate may be estimated roughly from the potential gradient measured at the concrete surface combined with concrete resistivity [4,8].

1.2. Corrosion Rates

Quantitative information on the corrosion rate of steel in concrete is of great importance for the evaluation of repair methods in the laboratory, for service life prediction and structural assessment of corroding structures as well as for control of repair work on-site. When speaking about 'corrosion rate' of steel in concrete, two different meanings, average corrosion rate and instantaneous corrosion rate, have to be distinguished:

- The average corrosion rate is the 'engineering' value needed to input into models for lifetime calculations or to predict the development of structural degradation. It can be determined as an average value over a long period of time by measuring weight loss (valid strictly only for homogeneous corrosion, possible only in the laboratory) or loss in cross section of the steel (on-site). If the time of depassivation, i.e. the start of corrosion, is not known, as is usually the case, the calculated average corrosion rates will be underestimated. On real structures exposed to changing environmental conditions the average value is the integral over periods with low and high corrosion rates (Fig. 1).

- The instantaneous corrosion rate i_{corr} can be calculated from the polarisation resistance R_p, determined by stationary or non-stationary electrochemical methods. This calculation is straightforward and correct only for homogeneous current distribution and general corrosion, the implications for measurements on real structures and on locally corroding rebars are discussed below. From Fig. 1 it is clear that the comparison of 'corrosion rate' values measured at different times in the life of a structure or on different structures have to be corrected for the influence of moisture and temperature.

1.3. Determination of the Polarisation Resistance R_p

Most corrosion rate measurements of steel in concrete are based on the relation between the polarisation resistance R_p and the corrosion rate by the formula of Stern-

(a)

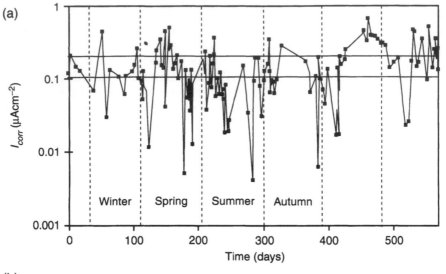

(b)

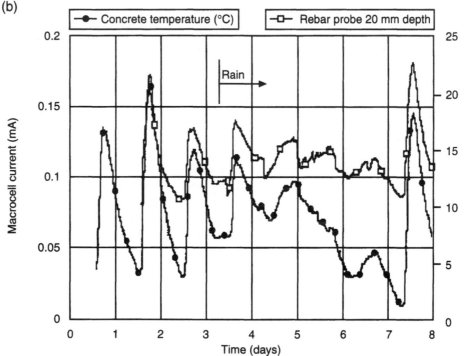

Fig. 1 *(a) Seasonal changes of* I_{corr} *values (determined from single polarisation resistance measurements) on a large beam exposed to a Madrid climate [27]. (b) Daily changes of the concrete temperature (depth 30 mm) and the macrocell current between an isolated, active steel and the rebars [28].*

Geary [9], $I_{corr} = B/R_p$, where B is a constant containing the anodic and cathodic Tafel slope. For actively corroding steel in concrete usually $B = 26$ mV is taken, for passive steel $B = 52$ mV [5, 10]. R_p and thus I_{corr} are related to the area of the sample under test, resulting in the specific polarisation resistance R_p^* (ohm cm²) and corrosion current density i_{corr} (μA cm^{-2}), called corrosion rate. 'True' corrosion rates are obtained only when the condition of linearity and steady state are fulfilled:

- Linearity usually is obtained by reducing the perturbation signal such that the polarisation of the electrode is limited to < 10 – 20 mV.

- Steady state implies long waiting times (theoretically $t \to \infty$) in potentiostatic tests, very low sweep rates (theoretically $dU/dt \to 0$) in potentiodynamic tests or frequencies $f \to 0$ in impedance measurements. None of these is possible in practice. Depending on the time constant of the system under test, finite waiting times or sweep rates can be used without further changes in the determined polarisation resistance [10]. On the other hand, long waiting times can lead to changes in the local environment at the steel/concrete interface that can disturb the measurements [11].

A further complication arises due to the presence of the ohmic or electrolyte resistance R_Ω (proportional to the electrolyte resistivity and to the distance between the reference electrode and the steel). Due to the high specific resistivity of mortar or concrete, values of R_Ω similar in magnitude to R_p can occur. R_Ω will be included in all the experimentally determined $R_{p,eff}$ values unless determined separately and subtracted. The consequence of not considering R_Ω is that values of R_p will be too high, and the corrosion rate i_{corr} underestimated. The error depends on the ratio R_Ω/R_p and not on the absolute magnitude of R_Ω.

Good correlation between the electrochemical weight loss, calculated by integration of R_p data from LPR measurements, and gravimetric measurements has been found [10]. Several comparative studies have shown that on small size laboratory samples [4,10,12–15] and/or on test blocks on-site with well defined current distribution [12,16] good agreement between the results of different measuring techniques, e.g. between electrochemical impedance (EIS) and galvanostatic pulse technique (GPM) measurements [14] as well as between GPM and LPR data [15] was obtained when the experimentally measured $R_{p,eff}$ values were corrected for the ohmic resistance. The largest differences occurred for passive reinforcement where non-steady state techniques such as the galvanostatic pulse technique or electrochemical impedance indicated somewhat lower R_p values.

2. Corrosion Rate Measurements On-Site

The main difference in measuring the polarisation resistance R_p not in the laboratory but on-site is the geometrical arrangement of the electrodes (Fig. 2): thus in the laboratory, samples are usually small and a homogeneous geometry can be achieved whereas on-site a non-uniform current distribution between the small counter electrode (CE) on the concrete surface and the rebar network (WE) will result.

2.1. Current Distribution

On real structures the area of the counter electrode is much smaller than that of the working electrode (rebars) and the electrical signal vanishes with increasing distance

(a)

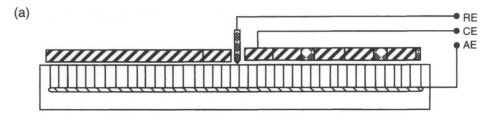

(b)

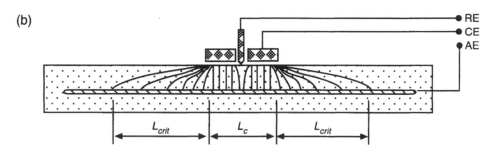

Fig. 2 The geometrical arrangement of the working electrode (AE) and the counter electrode (CE) determines the electrical field. (a) homogeneous current distribution, (b) non-homogeneous current distribution with current spreadout.

from the counter electrode (CE). As has been pointed out in the pioneering work of Feliù and Andrade [17,18] based on the transmission line approach, the measured (effective) polarisation resistance $R_{p,eff}$ is related to an unknown rebar surface area and cannot be converted directly to a corrosion rate by the Stern-Geary equation. A measure of the current spreadout is the critical length L_{crit} defined as that distance where the current has dropped to less than 10%. L_{crit} depends both on the concrete resistivity and on the specific polarisation resistance $R_p{}^*$ of the rebar — and thus on the corrosion state. Literature results [17–19], experiments with different CE on-site [19] and calculations with an electrical network [20] taking into account the geometrical arrangement, concrete resistivity and corrosion state (active or passive) of the rebars allow the following conclusions to be reached.

- **Actively corroding rebars:** in the case of homogeneous, actively corroding rebars, i.e. with small specific polarisation resistance $R_p{}^*$, the applied current is concentrated almost completely beneath the counter electrode (Fig. 3). For a concrete cover of 3 cm and a diameter of the counter electrode of 14 cm ($L = 7$ cm) the value of L_{crit} is calculated to about 1 cm. Higher cover results in an increase and higher concrete resistivities in a decrease of the critical length. Thus, it can be concluded that in the case of actively corroding rebars the current applied from a small CE on the concrete surface tends to confine itself and correct values of R_p are measured without signal confinement. This has been confirmed recently on-site, in which measurements were conducted with two GPM devices with different CE sizes and the same specific polarisation resistance was obtained [21].

- **Passive rebars:** passive steel in concrete has a very high specific polarisation resistance leading to a large current spreadout (Figs 2, 3). Calculations have shown that L_{crit} can be as high as 40 cm in low resistive concrete. The measured effective polarisation resistance $R_{p,eff}$ is thus related to an area of rebars that is up to 100 times larger than the CE surface. Several methods to determine correct R_p values in structures with passive rebars have been proposed [12,19 and literature cited therein]. The concept of a sensorised guard ring [22] avoids the current spreadout on passive rebars.

- **Localised corrosion:** in the case of local corrosion, small actively corroding areas are coupled to large passive areas. The excitation current coming from the counter electrode on the concrete surface enters predominantly the active areas (hot spots) with low specific R_p* (Fig. 3). This has been shown both by calculations and by laboratory experiments on model-macrocells [23] (Fig. 4). This current concentration (self confinement) is more pronounced with increasing cover depth and decreasing size of the counter electrode. It cannot be avoided by using a guard ring.

	Active	Passive	Localised attack
Electrical field for d.c. current	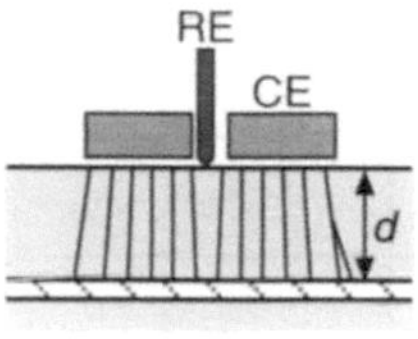	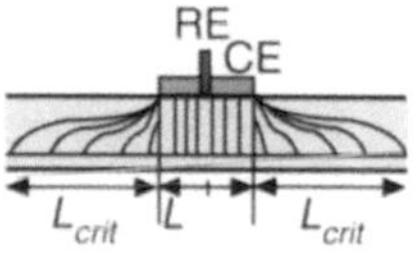	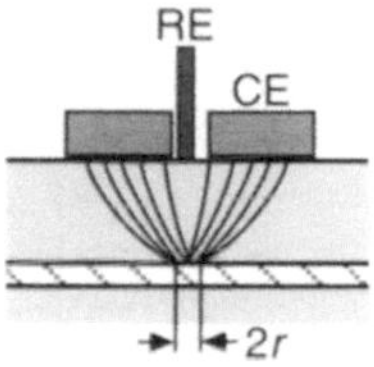
Measured value	$R_{p,eff}$ active	$R_{p,eff}$ passive	$R_{p,eff}$ local
Current spreadout	$L/d > 3$	$L/d = 2.5$	$L/d = 2.5$, $d = 3$ cm concentration of the current
Concrete wet	L_{crit} ca. 1 cm	L_{crit} ca. 40 cm	
Concrete dry	$L_{crit} < 1$ cm	$L_{crit} < 15$ cm	
Specific Rp calculated with CE area A	$R_p = R_p$, active*A	Much too small	Too high
Corrosion rate	Correct	Much too high	Too low

Fig. 3 Current distribution and determination of the instantaneous corrosion rate from polarisation resistance measurements for the three limiting cases: homogeneous active, homogeneous passive and localised attack. No guard ring. **d**: *concrete cover;* **L**: *diameter of counter electrode;* **L**$_{crit:}$ *critical length of current spreadout*

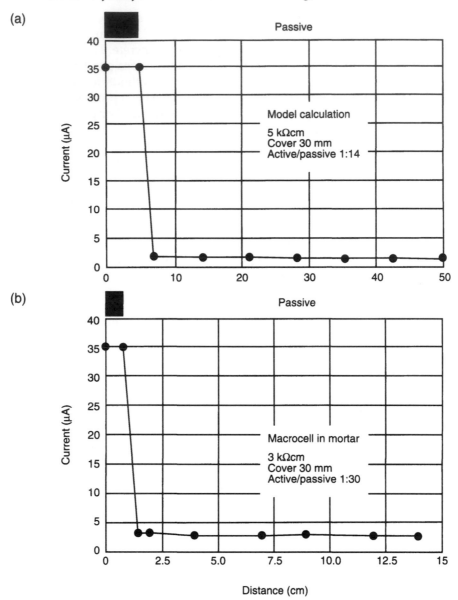

Fig. 4 *Current spreadout in the case of localised corrosion: (a) Simulation calculation and (b) experimental data from model macrocells [4,23].*

2.2. Determination of the Ohmic Resistance

As in laboratory experiments, the experimentally measured polarisation resistance $R_{p,eff}$ includes always the ohmic resistance R_Ω between the rebar network and the position of the reference electrode at the concrete surface. In order to obtain correct values for R_p the value of R_Ω has to be determined independently and subtracted from the experimentally measured polarisation resistance. The ohmic resistance can be measured either by high frequency a.c. signals, by imposing a short current pulse [4,12] or by current interruptor techniques. Both approaches, measuring at high frequencies or at short times, correspond to a high frequency current distribution [24]. The following are of importance:

- the steel/concrete interface at high frequencies can be represented by a (double layer) capacitance c_d, the corresponding impedance $Zc = 1/2\pi fc_d$ at high frequencies (f) is very small (in the order of ohms).

- With regard to the a.c. current distribution this corresponds to the situation where all the current is concentrated beneath the CE and no current spreadout occurs.

- The ohmic resistance determined corresponds always (independent of the corrosion state of the rebars) to the primary current distribution between the CE on the concrete surface and the rebars beneath the CE, identical to the current distribution for the case 'active rebars' (Fig. 3). Experimental confirmation of this fact is given by (a) the good proportionality found with concrete resistivities measured by the Wenner 4-point technique from the concrete surface [4,25] and (b) the experimental study performed on-site where the same specific resistivity of the concrete was measured using devices with different CE size [21].

In contrast, the measured values of $R_{p,eff}$ correspond to low frequency (ideally d.c. limit) current distribution. This difference in a.c. current distribution (for R_Ω) and d.c. current distribution (for $R_{p,eff}$) result in errors in the calculation of R_p when subtracting the ohmic resistance from the measured polarisation resistance. Thus:

- On passive rebars, the measured R_Ω will be too high compared to the area reached by the d.c. current and the error will increase with decreasing concrete resistivity.

- For locally corroding rebars, the measured R_Ω will be too low because the d.c. current is concentrated on the actively corroding area (Fig. 3).

- The measured R_Ω will be correct only for active general corrosion and can be subtracted from the d.c. value of $R_{p,eff}$.

Whereas signal confinement with a guard ring eliminates this problem on passive rebars (the area reached by high frequency and d.c. signals is the same), in the case of localised corrosion the underestimation of R_Ω results in R_p values that are too high and the instantaneous corrosion current I_{corr} results too small. These errors — an intrinsic problem of R_p measurements — are more pronounced for small localised corrosion areas (Fig. 5) as occur in chloride contaminated, humid concrete. The magnitude of these errors are of the same order as the uncertainties in the value of the B constant (about a factor of two).

Fig. 5 *Typical chloride induced localised corrosion on a rebar of a bridge deck.*

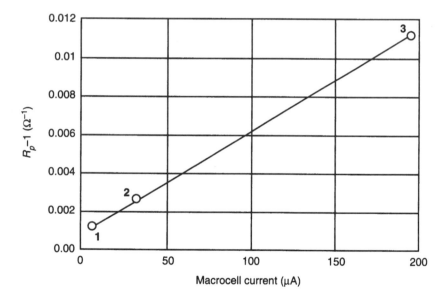

Fig. 6 *Reciprocal polarisation resistance, R_p^{-1}, determined from galvanostatic pulse measurements vs macrocell current on model macrocells in tap water (3), distilled water (2) and mortar (1) [23].*

2.3. Determination of Localised Corrosion Rates

The polarisation resistance determined on locally corroding rebars is related to the actively corroding area of the rebars, as has been shown by experiments on model macrocells where a linear proportionality between the corrosion current in the active/passive macrocell and the polarisation resistance measured has been found (Fig. 6) [23]. In a practical situation such as chloride induced localised corrosion of rebars in concrete, the area of the localised attack is not known and the local corrosion rate (penetration rate) cannot be calculated. This fact is crucial for life time predictions in frequent cases where loss in cross section and not concrete spalling is the critical factor for structural behaviour. Two ways to overcome this problem are proposed.

- The polarisation resistance R_p is related to the total geometrical area of the rebars beneath the counter electrode and an average corrosion current density is calculated with the Stern-Geary equation. Although mathematically correct, these values have no physical meaning because they are mixed values between a large area with $i_{corr} = 0$ (passive) and a small area with very high i_{corr} (localised attack). The inclusion of a 'pitting factor' improves the calculation but remains somehow arbitrary. Factors between 4 and 8 are reported for steel in concrete [19,26], the examination of pit depths on old RC structures lead to values greater than 10 [12] and experiments with model macrocells have resulted in factors as high as 15 [23].

- The experimental determination of the current distribution at the counter electrode using a segmented counter electrode [23] could be a promising way to determine the locally corroding area beneath the CE. Model experiments [23] have shown the feasibility in principal of the technique but a lot of research work has to be carried out in order to calculate the actively corroding area from the measured current distribution at the CE-segments.

2.4. Variation of i_{corr} with Time

Measuring the instantaneous corrosion current density from the concrete surface [5,27] or recording macrocell currents in specially instrumented bars or RC structures [28,29] allows the daily and seasonal variations of the instantaneous corrosion rate with temperature and humidity to be followed (Fig. 1). Although some proportionality between macrocell current and temperature has been found [27–29], the reasons for this behaviour in terms of changes in concrete resistivity, anodic and / or cathodic partial reaction of the corrosion process are not yet clear. In considering the use of on-site 'corrosion rate' measurements for service life prediction, control of the effectiveness of a repair work etc. it becomes clear that because of the huge fluctuations in instantaneous corrosion rate a simple comparison of two single measurements taken at different days with different temperature, humidity etc. can be highly misleading! Further work with a statistical evaluation of the whole measured set of data in order to extract mean and maximum corrosion rates is necessary.

3. Conclusions

It has been shown that measuring 'corrosion rate' of reinforcing steel on RC structures — despite the great need for meaningful life time predictions, control of effectiveness of repair work, etc. — is still a difficult task requiring specialists for the interpretation. Determination of the instantaneous corrosion rate is based on measurement of the polarisation resistance R_p, and the following are important:

- On homogeneous actively corroding rebars no or only negligible current spreadout occurs, and so the measured $R_{p,eff}$ values after correction for R_Ω can be related to the rebar area beneath the CE.
- On homogeneous passive rebars a large current spreadout has to be expected, the measured $R_{p,eff}$ values can be related to the area under the CE only when a guard ring for signal confinement is used.

- On locally corroding rebars the current is self confined to the actively corroding area and thus (when the actively corroding site has been located) no guard ring is needed. The ohmic resistance determined by high frequency a.c. signals or by current pulses corresponds to the total rebar area beneath the CE and underestimates the true R_Ω related to the actively corroding area. The measured value of $R_{p,eff}$ will relate to the (unknown) corroding area.

Calculation of instantaneous corrosion rates i_{corr} using the Stern-Geary equation is correct only in the case of homogeneous active rebars or on passive rebars when a guard ring is used. Especially in the frequent cases of chloride induced localised corrosion it is preferable to express the measurement results by the polarisation resistance $R_{p,eff}$. The problems with R_Ω compensation have to be taken into account with or without the use of a guard ring.

More fundamental studies on the influence of temperature on resistivity, on anodic and cathodic reactions of the corrosion process are needed in order to rationalise the dependence of the instantaneous corrosion rate on daily and seasonal temperature and humidity variations.

4. Glossary

R_p^* Specific polarisation resistance, area related, units ohm cm². this value can be used to calculate the corrosion current density i_{corr}.

R_p Polarisation resistance obtained after subtraction of R_Ω from the measured polarisation resistance $R_{p,eff}$, units ohm.

$R_{p,eff}$ Measured effective polarisation resistance, unit ohm. Not corrected for R_Ω and for the effects of current spreadout on large, passive rebar networks.

References

1. J. R. Stratfull, *Corrosion NACE*, 1957, **13**, 173t.
2. J. R. Van Daveer, *J. Am. Concr. Inst.* 1975, **12**, 697.
3. B. Elsener and H. Böhni, *Schweiz. Ing. Archit.*, 1987, **105**, 528.
4. B. Elsener and H. Böhni, "Potential mapping and corrosion of steel in concrete", in *Corrosion Rates of Steel in Concrete*, Eds N. S. Berke, V. Chaker and D. Whiting. ASTM STP 1065, American Society for Testing and Materials, West Conshohocken, Pa., 1990, p. 143–156.
5. B. Elsener, D. Flückiger, H. Wojtas and H. Böhni, Methoden zur Erfassung der Korrosion von Stahl in Beton, VSS Forschungsbericht Nr. 520, Verein Schweiz. Strassenachleute Zürich (1996).

6. B. Elsener, L. Zimmermann, D. Bürchler and H. Böhni, this volume, p. 125–140.

7. B. Elsener and H. Böhni, *Mater. Sci. Forum*, 1992, **111/112**, 635.

8. C. C. Naish, A. Harker and R. F. Carney, in *Corrosion of Reinforcement in Concrete*, Eds C. L. Page, K. W. Treadaway and P. B. Bamforth. Elsevier Applied Science London, 1990, p. 314–332.

9. M. Stern and A. L. Geary, *J. Electrochem. Soc.* 1957, **104**, 56.

10. C. Andrade, V. Castelo, C. Alonso and J.A. Gonzales, in Corrosion Effect of Stray Currents and the Techniques for Evaluating Corrosion of Rebars in Concrete, ASTM STP 906, Ed. V. Chaker, West Conshohocken, Pa., 1986, 43–63.

11. K. Videm, this volume p. 104–121.

12. B. Elsener, *Mater. Sci. Forum*, 1995, **192–194**, 857– 866.

13. A. Seghal, Y. T. Kho, K.Osseo-Asare and H. W. Pickering, *Corrosion NACE* 199, **48**, 871.

14. B. Elsener, H. Wojtas and H. Böhni, in *Corrosion and Corrosion Protection of Steel in Concrete*, Ed. R. Swamy. Sheffield Academic Press (1994), Vol. 1 p.236–246.

15. K. Menzel, D. Sonnentag and F. Paul, COST 509 D3, Report 1994.

16. R. Polder *et al.*, in *Corrosion and Corrosion Protection of Steel in Concrete*, Ed. R. Swamy. Sheffield Academic Press 1994, Vol. 1, p. 571–580.

17. S. Feliu, J. A. Gonzales, C. Andrade and V. Feliu, *Corrosion '87*, Paper 145, NACE, Houston, Tx., 1987.

18. S. Feliu, J. A. Gonzales, C. Andrade and V. Feliu, *Corrosion NACE*, 1988, **44**,7 61.

19. S. Feliu, J. A. Gonzales and C. Andrade, in *Techniques to Assess the Corrosion Activity of Steel Reinforced Concrete Structures*, eds N. S. Berke, E. Escalante, C. K. Nmai and D. Whiting. ASTM STP 1276, American Society for Testing and Materials West Conshohocken, Pa., 1996, 107.

20. B. Elsener, H. Wojtas and H. Böhni, Inspection and Monitoring of Reinforced Concrete Structures — Electrochemical Methods to Detect Corrosion, in *Proc. 12th Int. Corrosion Congr.*, NACE, Houston 1993, Vol 5A, p. 3260–3270.

21. B. Elsener, O. Klinghoffer, T. Frolund, E. Rislund, Y. Schiegg and H. Böhni, in *Proc. Int. Conf. Repair of Concrete Structures*, Svolvaer Norway, Ed. A. Blankvoll. Norwegian Public Roads Administration(1997 p. 391–400.

22. J. P. Broomfield, J. Rodriguez, L. M. Ortega and A. M. Garcia, "Corrosion Rate Measurements in Reinforced Concrete Structures by a Linear Polarization Device", *Int. Symp. on Condition Assessment, Protection Repair and Rehabilitation of Concrete Bridges Exposed to Aggressive Environments*, ACI Fall Convention, Minneapolis, 1993.

23. B. Elsener, A. Hug, D. Bürchler and H. Böhni, Evaluation of localised corrosion rate on steel in concrete by galvanostatic pulse technique, in *Corrosion of Reinforcement in Concrete Construction*, Eds C. L. Page, P. Bamforth and J. W. Figg. Soc. Chem. Ind., London, 1996, p. 264.

24. S. C. Kranc and A. A. Sagues, *Electrochim. Acta* 1993, **38**, 2055.

25. B. Elsener, Korrosion in Stahlbetontragwerken — Einflussgrössen und Möglichkeiten der Ueberwachung, in *Erhaltung von Brücken*, SIA Dokumentation D099, Schweiz. Ing. und Arch.verein, Zürich 1993, p. 45.

26. J. A. Gonzales, C. Andrade, C. Alonso and S. Feliu, *Cement and Concrete Res.*, 1995, **25**, 257.

27. C. Andrade, J. Sarria and C. Alonso, Statistical study of the simultaneous monitoring of rebar corrosion rate and internal humidity in concrete structures exposed to atmosphere, in *Corrosion of Reinforcement in Concrete Construction*, Eds C. L. Page, P. Bamforth and J. W. Figg. Soc. Chem. Ind., 1996, p. 233.

28. L. Zimmermann, Y. Schiegg, B. Elsener and H. Böhni, Electrochemical techniques for monitoring the conditions of concrete bridge structures. *Proc. Int. Conf. Repair of Concrete Structures*, Ed. A. Blankvoll. Norwegian Public Roads Administration, Svolvær, Norway, 1997, p. 213.

29. B. Elsener, *Schweiz. Ing. Archit.*, 1997, **115**, 4.

10

Field and Laboratory Experience with Potentiostatic Polarisation and Potentiokinetic Scans to Assess the Severity of Corrosion of Steel in Concrete

K. VIDEM

Center for Materials Science, University of Oslo, Gaustadalleen 21, N-0371 Oslo, Norway

ABSTRACT

Laboratory experiments have been carried out in synthetic pore water solutions and in mortar slabs with and without chloride additions. Field studies were performed at a 840 m long, 15 year old bridge at the coast of Northern Norway. The corrosion film on steel in concrete is changed by polarisation. The impressed current also induces changes of the pore water in contact with the metal. This leads to a complicated behaviour under polarisation and a large pseudo-capacitance. Potentiodynamic scans could not be used to determine the linear polarisation resistance of the cut reinforcement bars in the 15 year old bridge due to the large pseudo-capacitance of steel in concrete, giving a charging current that overshadowed the Faraday current. Polarisation resistance obtained by potentiostatic exposures varied more than a decade depending upon the polarisation time. Nevertheless, this technique discriminates between sites with high and low corrosion rates.

1. Introduction

Potential mapping according to ASTM C876-91 [1] is the most commonly applied electrochemical technique to assess corrosion of steel reinforcement in concrete, as it is simple and well established. It is generally accepted that potential mapping must be supplemented with other methods. However, more expensive and time consuming electrochemical techniques are only justified if they give more reliable guidance. The situation is not fully satisfactory, as it is more difficult to apply electrochemical techniques to asses the corrosion of steel in concrete than for metals exposed to aqueous solutions. The use of polarisation data to calculate corrosion rates requires that the mechanism of the corrosion process is sufficiently well understood. For the simplest condition with activation control of both anodic and cathodic reactions, the corrosion rate can be obtained from the linear polarisation resistance by the Stern-Geary equation [2]. This is not the case for steel in concrete. However, for many combinations of metal and environment the corrosion rate can be estimated from the polarisation resistance also when the conditions for the Stern-Geary equation are not fulfilled, provided that sufficient practical experience is available. However, this is not simple for steel in concrete. The electrochemistry is complex and not well understood. Reliable Tafel slopes have not been determined, and the data base for relating corrosion rates to polarisation resistance is minimal.

The linear polarisation resistance can be measured by various methods — potentiodynamic scans, potentiostatic polarisation, galvanostatic pulses, cyclic voltammetry and a.c. impedance spectroscopy — all having been studied by the present author [3–5]. As will be shown, the linear polarisation resistance can vary by more than an order of magnitude for a given case with steel in concrete, depending upon the type of method applied and the test parameters used. Obviously, a general relationship between linear polarisation resistance and the corrosion rate does not exist for steel in concrete. Thus, the 'constant' used to relate corrosion rates to linear polarisation resistance must be suitable for the specific test method. The present studies in the laboratory and in the field examine the reliability of potentiodynamic scans and potentiostatic polarisation for estimating the corrosion rate of steel in concrete.

2. Experimental Technique

The experimental techniques for the laboratory experiments are described in recent papers [3–5]. Carbon steel was exposed in 0.1M KOH solutions with different concentrations of NaCl. Specimens were cut from 1 mm cold rolled steel sheet (C 0.07, Si 0.02, Mn 0.029). A PVC insulated lead for electrical connection was soldered to the specimens and the connecting point and the upper region of the specimens masked off with epoxy. The last step in the surface treatment of the steel was pickling for 3 min in 10 wt% HCl. The same types of steel electrodes were cast into mortar slabs $60 \times 60 \times 60$ mm. Mortar with a water/cement ratio of 0.6 and sand/Portland cement ratio 3:1 were made with and without NaCl. These slabs were cast 11 months before the measurements reported. Reference electrodes were placed on the external surface during the measurements. Electrochemical experiments were carried out with conventional equipment.

The field studies were carried out at the 840 m long, 15 year old Gimsøystraumen bridge at the coast of Northern Norway [6]. Three years before the reported measurements, electrode systems for electrochemical studies were arranged by cutting reinforcement bars and installing reference electrodes and counter electrodes.

3. Results

3.1. Potentiodynamic Scans and Cyclic Voltammetry

Figure 1 shows the current as a function of potential for cyclic voltammetry of a Randles circuit. As indicated on the figure, the sum of the electrolyte resistance (between the reference and working electrode) and the linear polarisation resistance R_{lpr} can be identified from the slope of the plot. The slope is independent of scan rate in this case. The capacitance can be calculated from the distance between the two lines. This distance is proportional to the scan rate. Figure 2 shows an example of cyclic voltammetry for steel exposed two months in 0.1M KOH solution with scans of 15 mV to both sides of the free corrosion potential. The behaviour of this electrode

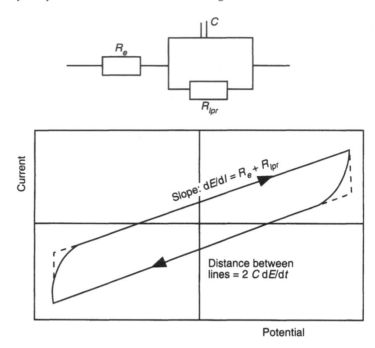

Fig. 1 *Principle diagram for the current density as a function of difference from corrosion potential for cyclic voltammetry. The dotted lines apply for the electrolyte resistance R_e being zero.*

was in reasonable harmony with a Randles circuit with a small value of the series resistor. The two tangents to the curves when potential passes the free corrosion potential (one at rising and one at falling potential) give two values for the linear polarisation resistance. The polarisation resistance was nearly the same for the positive and negative scan direction in aerated KOH solutions without chloride and was little influenced by the scan rate. In addition to the Faraday current of the electrochemical reactions, current is also required for charging of the interfacial capacitance of the electrode. The capacitance values determined from this varied only a little with the scan rate.

Table 1 sums up polarisation resistance and capacitance values obtained by cyclic voltammetry of steel exposed two months in 0.1M KOH and in 0.1M KOH + 0.05M NaCl. It is seen that in the chloride-containing solution capacitance values were far from constant, but decreased with increasing scan rates.

Figure 3 shows curves for steel in cement mortar without chloride added. In contrast to Fig. 2, the change of scan direction had only a small instantaneous effect on the current density. Similar experiments with mortar slabs with chloride added as 3% of the cement weight resulted in curves with about the same shape, but higher current densities, as shown in Fig. 4. Table 2 shows values for polarisation resistance and capacitance obtained by potentiodynamic scans with different rates for steel in the mortar slabs. It is seen that the value for polarisation resistance for the steel in chloride free mortar decreased by a factor of two with increase of the scanning rate from 0.1 to 1 mVs^{-1}. The effect of scanning rate for mortar with chloride was slightly

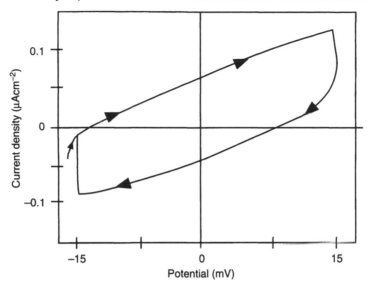

Fig. 2 *Current density as a function of difference from corrosion potential for cyclic voltammetry in 0.1M KOH at a rate of 1 mVs^{-1}; 2 months exposure.*

less. The capacitance obtained by cyclic voltammetry was more affected by the scanning rate than the polarisation resistance.

When Tables 1 and 2 are compared, it is noted that addition of chloride to the 0.1M KOH solution decreased the polarisation resistance more than an order of magnitude, while in mortar the decrease caused by chloride was only a factor of about three.

Table 1. *Values of linear polarisation resistance and capacitance of steel in solutions measured by cyclic voltammetry with different scan rates*

Scan rate (mVs^{-1})	Linear polarisation resistance (kΩcm^2)		Capacitance (μFcm^{-2})	
	0.1M KOH	0.1M KOH + 0.05M NaCl	0.1M KOH	0.1M KOH + 0.05M NaCl
−0.1	300	23.4		
0.1	300	23.4	27	2133
−0.4	300	13.8		
0.4	300	14.1	20	1066
−1.0	228	8.7		
1.0	255	9.0	28	760

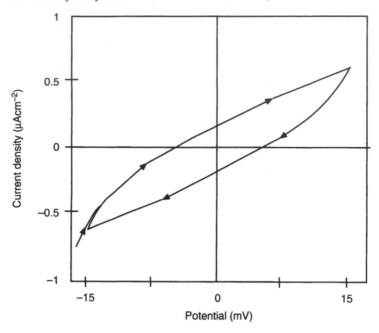

Fig. 3 *Current density as a function of difference from corrosion potential at cyclic voltammetry with steel 11 months in cement mortar without chloride. Scan rate 1 mVs⁻¹.*

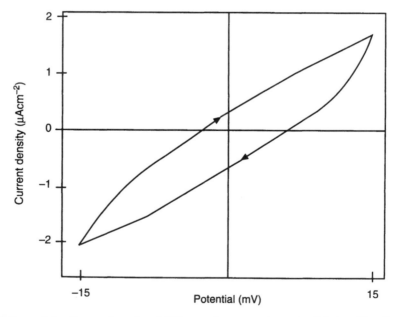

Fig. 4 *Current density as a function of difference from corrosion potential at cyclic voltammetry with steel 11 months in cement mortar with 3 % chloride of the cement weight. Scan rate 1 mVs⁻¹.*

Test 'cells' with 40 cm² cut reinforcement bars, embedded reference electrodes and counter electrodes were available in 14 locations at Gimsøystraumen bridge [6].

Table 2. *Values of linear polarisation resistance and capacitance of steel in cement mortar measured by cyclic voltammetry with different scan rates.*

Scan rate (mVs⁻¹)	Linear polarisation resistance (kΩcm²)		Capacitance (μFcm⁻²)	
	Mortar no Cl⁻	Mortar 3% Cl⁻	Mortar no Cl⁻	Mortar 3% Cl⁻
−0.1	63	23.4		
0.1	66	23.4	1173	2466
−0.4	38	14.8		
0.4	41	18.1	353	866
−1.0	29	12.3		
1.0	32	12.5	170	386

Originally a standardised procedure was used with scanning from −13 mV from the corrosion potential to +13 mV with a rate of 0.5 mVs⁻¹ [6]. It turned out to be difficult to use automatic compensation of the electrolyte resistance despite the reference electrodes being situated only about 1 cm from the cut reinforcement bars. This compensation with an analogue potentiostat powered from the 230 V mains led to a high noise level and sometimes free oscillations. It was hoped that curves without compensation for the electrolyte resistance could be sufficient for detecting trends in the corrosion rate. However, changes in the electrolyte resistance had too large an effect. Figure 5 shows an example of the results when compensation for the electrolyte resistance was not applied. Often the curves had a shape that hardly could be used to calculate any linear polarisation resistance.

As the current for charging a capacitance is reduced when the rate of potential change is reduced, experiments with different scanning rates were carried out with the aim of improving the situation. A potentiostat with digital control was used. The noise level was much lower with battery operation than with power from the mains. Only in the latter cases was use of compensation of the electrolyte resistance possible. The slower the scan rate, the higher was the apparent polarisation resistance, in agreement with the observations with laboratory cement mortar slabs. However, consistent and useful results were not obtained even with scan rates as low as 0.02 mVs⁻¹.

3.2. Potentiostatic Polarisation

Figure 6 shows the current density as a function of time for potentiostatic polarisation 20 mV above the free corrosion for steel exposed in 0.1M KOH, in 0.1M KOH + 0.05M NaCl and in cement mortar with chloride mixed in as 3% of the cement weight. The trend was for the current density to decrease with time for a period. For 0.1M KOH + 0.05M NaCl an increase in current density occurred after 20 min. This happened because the anodic polarisation accelerated localised attack. The effect of chloride in mortar is illustrated in Fig. 7. It is seen that the current density decreased with time, but tended to stabilse after about 2 h. The polarisation resistance increased with time accordingly, as shown in Fig. 8.

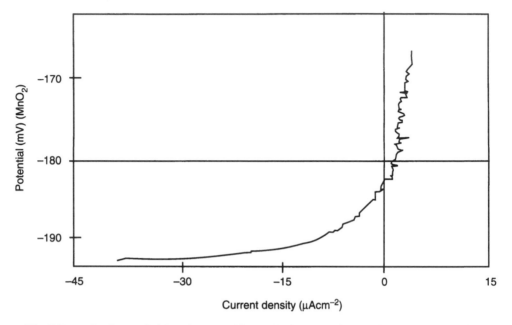

Fig. 5 *Example of potentiodynamic scan with cut reinforcement bars in the 15-year old bridge.*

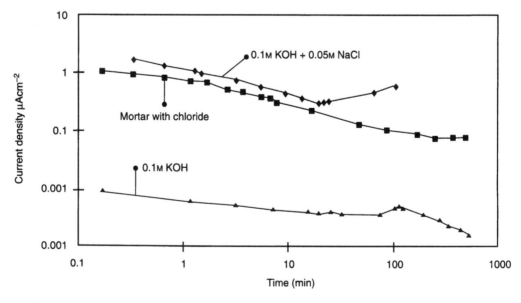

Fig. 6 *Current density as a function of time for steel polarised under potentiostatic conditions 20 mV above the corrosion potential in 0.1M KOH, in 0.1M KOH + 0.05M NaCl and in cement mortar with 3 % chloride of the cement weight.*

Table 3 shows values for polarisation resistance obtained by potentiostatic polarisation 20 mV above the free corrosion potential for the same tests treated in Table 1 and 2 with cyclic voltammetry. The corrosion potentials are given in Table 4.

For the steel with 15 years exposure in Gimsøystraumen bridge, the current density decreased with time for a very long period under potentiostatic polarisation. No approach to a steady state current was obtained for the longest testing time tried (11 h). Figures 9 and 10 show the current density as a function of time for a location

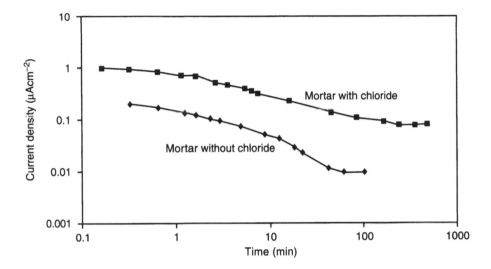

Fig. 7 *Current density as a function of time for steel polarised under potentiostatic conditions 20 mV above the corrosion potential in cement mortar without chloride and with 3% chloride of the cement weight.*

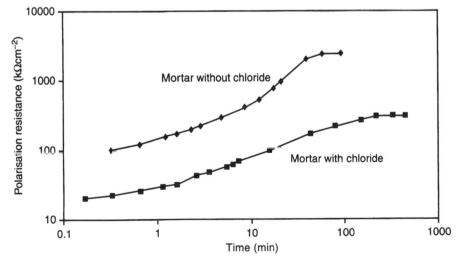

Fig. 8 *Polarisation resistance as a function of time for steel polarised under potentiostatic conditions 20 mV above the corrosion potential in cement mortar without chloride and with 3% chloride of the cement weight.*

Table 3. *Values of linear polarisation resistance of steel in different environments, measured by potentiostatic polarisation at different times.*

Time	Linear polarisation resistance (kΩcm^2)			
(s)	0.1M KOH	0.1M KOH + 0.05M NaCl	Mortar no Cl$^-$	Mortar 3% Cl$^-$
12	2000	8	80	20
100	3000	12	200	30
1000	4000			

Table 4. *Corrosion potentials.*

Environment	E, V (SCE)
Ca (OH)$_2$	–122
Mortar (no Cl$^-$)	–200
Mortar (3% Cl$^-$)	–501
Bridge, low rate, Figs, 9, 11	–190
Bridge, high rate Figs 10, 12	–327

with a very low and a moderate corrosion rate respectively. These plots were obtained without automatic compensation of the resistance drop between the reference electrode and the working electrode. The resistance value for calculating the potential drop was determined separately with a.c. at 1000 Hz and varied between 500 and 5000 VA^{-1}. Measurements at 100 Hz gave about 10% higher values. The polarisation resistance as a function of time is given in Figs 11 and 12. Both values with and without correction for the resistance drop are included. The polarising current at sites with low corrosion rates was so low that the contribution of the resistance drop was unimportant. This contribution was significant for the sites with the highest corrosion rates, despite the resistivity in the concrete being lower. As already stated, the distance between the reference electrode and the working electrode was only about 1 cm.

Additional numerical values for polarisation resistance from the field are given in Table 5. The lowest value in the field was of the same order of magnitude as for mortar with 3% Cl$^-$: The highest values in the field were much higher than obtained in the laboratory, reflecting the fact that steel may corrode very slowly in concrete after 15 years exposure.

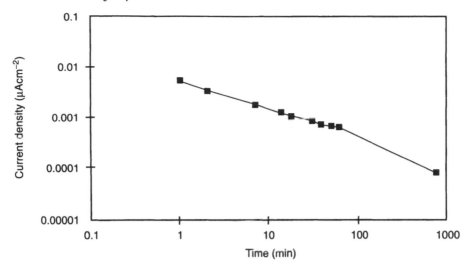

Fig. 9 *Current density as a function of time for 40 cm² cut reinforcement bar polarised under potentiostatic conditions. Site with a very low corrosion rate, 15 years exposure.*

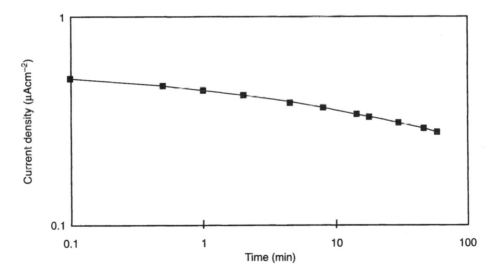

Fig. 10 *Current density as a function of time for 40 cm² cut reinforcement bar polarised under potentiostatic conditions. Site with a moderate corrosion rate, 15 years exposure.*

4. Discussion
4.1. Polarisation Resistance Obtained by Potentiodynamic Polarisation

Laboratory studies of the electrochemistry of steel in cement mortar show very large values for apparent interfacial capacitance. Up to 2466 μFcm^{-2} was obtained with cyclic voltammetry at 0.1 mVs^{-1} for steel in the mortar block with 3 % Cl$^-$. As described later, this very high capacitance does not originate from a physically existing interface

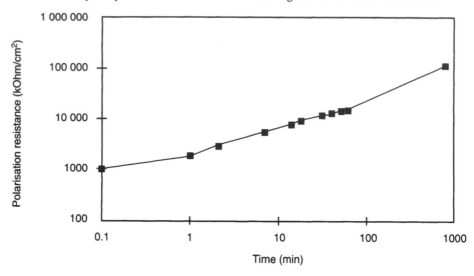

Fig. 11 *Polarisation resistance as a function of time for 40 cm² cut reinforcement bar polarised under potentiostatic conditions. Site with a very low corrosion rate, 15 years exposure.*

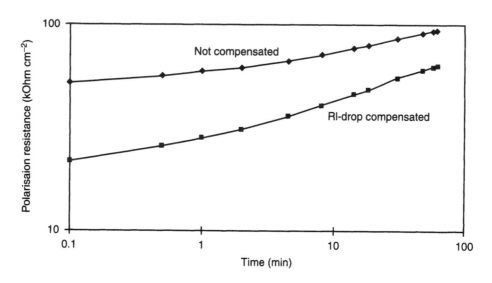

Fig. 12 *Polarisation resistance obtained with and without compensation of the resistance drop as a function of time for 40 cm² cut reinforcement bar polarised under potentiostatic conditions. Site with a moderate corrosion rate, 15 years exposure.*

capacitor. Nevertheless, charging currents are set up when the potential of the electrodes is changed. The charging current of the capacitance mentioned above with a scan rate of $0.5\,\mathrm{mVs^{-1}}$ is $1.3\,\mathrm{\mu Acm^{-2}}$ and so high that the Faraday current associated with corrosion tends to be overshadowed. Therefore, a problem with assessing severity of corrosion from potentiodynamic polarisation was not a surprise, despite this the technique is in practical use.

Table 5. *Examples of polarisation resistance values obtained at different locations in Gimsøystraumen bridge at different times with potentiostatic polarisation 20 mV above corrosion potential. 40 cm^2 cut reinforcement bars, 15 years exposure*

Location	Polarisation resistance (kΩ cm^2)		
	1 min	30 min	764 min
A	28	56	
B	830	3600	
C	332	800	
D	1700	11 000	114 000
E	436	3200	
F	284	1600	
G	124	532	

The difficulties with applying a potentiostat with automatic compensation of the resistance drop between the reference electrode and working electrode originate from the high apparent capacitance. The automatic compensation acts as a positive feedback. Increased capacitance of the electrode leads to increased amplification of the noise and reduced stability. Above a certain capacitance free oscillations take place. The current is higher in a potentiodynamic scan than in a potentiostatic exposure of some minutes duration or longer. Therefore, will the contribution of the resistance drop between the reference electrode and working electrode be higher in potentiodynamic tests? Corrections for this resistance drop appear to be a necessity in this case, and unfortunately more difficult to carry out. However, automatic compensation is not the only solution. Using a.c. measurements to determine resistance and take care of the resistance drop by computations after the finished measurements is an alternative — as sometimes used in this study.

The pseudo-capacitance is not a constant, but varies with the rate of potential change. The faster the potential change, the larger is the apparent capacitance, as seen from Tables 1 and 2. Thus, the steel electrode in concrete does not behave like a single Randles circuit. It is no longer strictly justified to obtain the linear polarisation resistance from the tangent to the curve at the free corrosion potential. For the steel electrodes with 15 years exposure the curves obtained with a scan rate of 0.5 mVs^{-1} could hardly be used to calculate any linear polarsation resistance.

It is well established that there is a limit to how fast potentiodynamic scans can be carried out for assessing the severity of corrosion. With high corrosion rates the charging current for the capacitance becomes relatively less important. Thus, it is obvious that problems introduced by high capacitance will be smaller in the case of a high corrosion rate. It is also obvious that higher scan rates are allowable for higher corrosion rates. To obtain better knowledge about this for the steel with 15 years exposure, experiments with different scanning rates were carried out. The slower the scan rate, the higher was the apparent polarisation resistance, in agreement with the observations with laboratory cement mortar slabs. Even scan rates as low as 0.02 mVs^{-1} did not eliminate this effect. A still more severe problem was that the polarisation resistance values obtained often were very different for a positive and a negative scan direction. No procedure was found that gave satisfactory results. The problems described applied for cut reinforcement bars with an area of 40 cm^2.

If the problems had originated only from a high capacitance, slow scans would have solved them. The results from potentiostatic experiments illustrated in Figs 8–12 and Tables 3 and 5 explain why this was not the case. Large time induced changes of the current density occurred under potentiostatic conditions. Therefore, the current density in tests with cyclic voltammetry and potentiodynamic scans will be a mixed effect of two variables: potential change and time. This type of data can be problematic to interpret for practical applications. Under these conditions electrochemical techniques that use a constant potential or a constant current are better than those with a varied potential (potentiodynamic scans, cyclic voltammetry and a.c. impedance).

The current in potentiostatic tests was reasonably stable after 0.5 h in the KOH solution and after about 2 h in cement mortar. Polarisation induced changes of the steel electrodes occurring with time were small in KOH solutions. However, for mortar slabs they were always disturbing, but less problematic when the slabs were relatively new. The experiments reported in this document were carried out 11 months after casting of the slabs. For the steel in the 15-year old bridge, polarisation changed the electrodes so much with time, that it was hardly possible to determine any reliable linear polarisation resistance by potentiodynamic scanning. It is natural to speculate on why this important aspect is not given more attention in the literature. Can it be because experiments with steel in concrete often have been carried out with 'young' laboratory specimens?

Electrochemical methods for determination of the corrosion rate are expected to give a rough indication rather than an exact number for steel in concrete. 'Be able to distinguish between active and passive corrosion', is a popular wording. From the present experience in the field it is questionable whether potentiodynamic scans give more reliable information about the severity of corrosion than an analysis based on corrosion potential and potential gradients. Attention should also be drawn to the problems with compensation of the resistance drop between the reference electrode and the counter electrode; if slow scanning is required, the measurements become expensive for routine applications.

Regarding the problems with noise and stable operation, it is emphasised that different potentiostats perform differently, as gain and time constants of some electronic circuits are important parameters. Laboratory potentiostats are designed for a fast response. Ability to accept high capacitance is more important for tests with steel in concrete. Potentiostats with adjustable time constants would be helpful [7].

The experience described above apply for the ideal case with electrodes with a defined area and with reference electrodes embedded in the concrete nearby. Experiments with polarisation of the main reinforcement were also carried out. These measurements without confinement of the current with a guard electrode could not be interpreted at all.

4.2. Polarisation Resistance Obtained by Potentiostatic Polarisation

The current for activation controlled reactions depends on potential and is independent of time. This is apparently not the case for steel in concrete. It can be seen from Fig. 9 that the values from potentiostatic tests varied by two decades depending on the polarisation time used. Values obtained by the potentiodynamic scans differ from those obtained by the potentiostatic technique. Thus, one can draw the important conclusion that a value for polarisation resistance alone is not a satisfactory base for estimation of the corrosion rate.

In the alkaline solutions a steady state current was reached after about 0.5 h with potentiostatic polarisation. For 11 month old mortar slabs it took about 2 h. For the steel specimens in the bridge no approach to a steady state was observed even for the longest measuring period tried (11 h). The current was found to decrease with time and followed roughly the relationship:

$$I = I_o \times t^{-\alpha} \tag{1}$$

where I = current, t – time and I_o and α = 'constants' that varied from test location to test location. α was close to 0.5 in most cases in the field, the exception being shown in Fig. 10. This case is regarded as chloride induced corrosion. It is not to be expected that the current vs time for chloride induced corrosion will follow a consistent pattern. Cathodic polarisation gives some cathodic protection that can stifle pitting, and anodic polarisation tends to initiate additional localised attack. Therefore, the decrease of current with time under anodic polarisation will be less pronounced for chloride contaminated concrete. The current can even increase with time as shown for 0.1M KOH + 0.05M NaCl in Fig. 6. However, the shapes of the curves from the various sites in the bridge were sufficiently similar to treat their differences with a single number I_o. High I_o indicates a high corrosion rate. A slow decrease of the current with time under anodic polarisation is also indicative of chloride induced corrosion. Today the ranking can be relative only, as a data base sufficient to relate I_o to corrosion rates is lacking. More work on this is encouraged as the experience with anodic, potentiostatic polarisation for assessment of the severity of corrosion is promising.

As the current density after some minutes potentiostatic polarisation tends to be low, compensation of the resistance drop between the reference electrode and the working electrode is less important than for potentiodynamic testing. As this compensation can be difficult, the potentiostatic technique has an advantage also in this respect. In the present work the contribution of the resistance drop was elaborated from a.c. measurements and computations after the measurements were carried out. Resistance in the concrete measured at 100 Hz was roughly 10% higher than at 1000 Hz, but this difference varied for different concrete qualities in another study. Concrete is not an Ohmic conductor as pointed out very clearly by Keddam *et al.* in a recent publication [8]. Therefore, an exact correction of the resistance drop is more complicated than normally considered. This is an additional argument for electrochemical techniques where this contribution is small.

4.3. The Origin of the Polarisation Induced Changes of the Electrodes

The nature of the changes that occur with time due to polarisation of steel is taken up in our earlier publications [3–6]. Anodic polarisation of only some seconds decreased the capacitance and cathodic polarisation increased it. Anodic oxide film growth is made responsible for the reduction in capacitance under anodic polarisation in chloride free, alkaline environments. The mechanism responsible for the increase of capacitance by cathodic polarisation is unclear, even though film dissolution is anticipated to contribute. It is difficult to study this by electron optical techniques, as it hardly possible to obtain undisturbed corrosion films from iron in cement mortar for investigation. The closest one can reach appears to be films formed in synthetic concrete pore water. Transmission electron diffraction of films formed in 0.1M KOH indicated a thin amorphous film. This, combined with the general chemistry of iron in the relevant pH and potential region indicate that polarisation can introduce modifications of the stoichiometry of the corrosion film in addition to changes in thickness. Such redox reactions consume current and result in changes in the properties of the film. Similar conclusions are drawn by Andrade *et al.* [9] and are used to explain the large pseudo-capacitance observed with a.c. at low frequencies. Redox reactions take place in a relatively narrow potential range. The corrosion potential of steel in concrete varies in a wide region (about –0.6 to 0 V (SCE) in the present experiments). It is then difficult to understand that a contribution from such reactions always appears to occur.

Another effect of polarisation originates from the very small amount of pore water in concrete in contact with the steel. Anodic polarisation reduces the pH of this water and cathodic polarisation increases it. Other changes in composition, for example the oxygen level at the steel surface also take place. These effects can be large because, there is little water. The response of electrodes influenced by diffusion of a single component in the environment is treated in electrochemical textbooks [10]. The steel in concrete system is much more complicated and can hardly be treated mathematically at present. It is difficult to formulate any opinion about which of the two processes contribute the most. The pore system and moisture level in concrete exhibit large variations. The rate of redox reactions introduced by polarisation will also vary, bearing in mind that exposure times are very different, the steel rusted when the structure was built, and that localised attack and passive corrosion are different processes.

Capacitance in electronic circuits is usually determined from the phase shift between current and potential. Both the transformations of the corrosion film and of the alteration of the environment take time, leading to a potential response that is retarded in phase with respect to the current. This phase shift resembles the action of a capacitance and creates the very large pseudo-capacitance. This capacitance is not related to a real capacitor. Its magnitude varies with the test conditions. The phase shift between potential and current also affects the apparent resistance in the system, so this apparent resistance becomes different from the Faraday polarisation resistance.

4.4. The Relationship Between Polarisation Resistance and Corrosion Rate

The Stern-Geary treatment [2] is based on knowledge of the anodic and cathodic Tafel slopes and the assumption that polarisation alters only the rate of these reactions.

This is not fulfilled for steel in concrete. Therefore, different views have been expressed about the constants to be used for transforming polarisation resistance to corrosion rate. Andrade *et al.* [11] suggest 20 mV for active corrosion and 60 mV for corrosion in the passive state. The differentiation is justified as one is faced with two different mechanisms of attack. Figure 11 shows a variation of the polarisation of two decades depending of the time for the potentiostatic polarisation. Potentiodynamic polarisation led to values influenced by the rate of potential change. Thus, it is obvious that a generally applicable constant for transforming polarisation resistance to corrosion rate does not exist for steel in concrete. It is important that there is harmony between the test technique and the 'constant' for calculation of the corrosion rate.

It is seen from Table 1 and 2 that addition of chloride to the 0.1M KOH solution decreased the polarisation resistance by more than an order of magnitude, while in mortar the decrease caused by chloride was only a factor of about three. It appears unreasonable that the corrosion rate for the mortar with 3% chloride of the cement weight was only three times higher that for mortar without chloride. This relatively small difference in corrosion rate is hardly reasonable from the corrosion potentials which were –200 and –501 mV(SCE) (Table 4) for the mortar without and with chloride respectively. Thus, there are reasons to question whether the polarisation resistance did reflect the difference in corrosion rate in this case, even though any definite conclusion cannot be drawn from the available data.

In an interesting publication Gonzalez *et al.* present a careful mathematical analysis of data from potentiodynamic polarisation and cyclic voltammetry based on a single Randles circuit [12]. Qualitatively the shape of the plots in Figs 2–4 appears to fit with the theory. However, from the time induced changes described above, it is understood that the electrodes with 15 years exposure in the bridge can hardly be represented by a single Randles circuit. One can simulate the behaviour of the steel electrode in concrete better by applying more than one Randles circuit. This is taken up in a recent report [3] and is explored in detail by Millard *et al.* [13] for a.c. impedance and galvanostatic pulse measurements. Up to five Randles circuits in series had to be applied. However, the question then arises, which of the resistance-capacitance RC-links are related to the corrosion rate? Millard *et al.* [13] disregarded some of the RC-links for the estimation of the corrosion rate due to too low or too high time constants. Reasonable values for the corrosion rate were obtained by this treatment.

Another approach presented by Sagues *et al.* [14] is to consider a constant phase angle element in the a.c. system. This was justified from a.c. impedance studies. Sagues *et al.* present a mathematical treatment that allows one to calculate a linear polarisation resistance from cyclic voltammetry or potentiodynamic scans with two scan rates. For certain, this is an interesting approach. However, the calculation of corrosion rates from the polarisation resistance obtained is not necessarily straightforward, as it is dependent upon an understanding of the phenomena responsible for the a.c. constant phase angle element. As described in textbooks of electrochemistry [10], diffusion of a single substance in the environment gives a phase angle of $p/4$ independent of the frequency for an electrode that is reversible for the diffusing substance. However, the polarising current creates effects that are too complex to allow a satisfactorily theoretical treatment at present. Attention is drawn to the fact that the effects described probably make it impossible to obtain electrochemically meaningful Tafel data for steel in old concrete by any direct method.

5. Conclusions

Anodic polarisation causes film growth and cathodic polarisation reduction of the film thickness. Possibly, polarisation also influences the stoichiometry of the corrosion film. Due to the small volume in pore water, polarisation may also alter the environment. These changes lead to a complex polarisation behaviour of steel in concrete. Therefore, some difficulties arise in assessing the severity of the corrosion of steel in concrete from electrochemical techniques with polarsation of long duration.

Both the transformations of the corrosion film and of the alteration of the environment take time, leading to a potential response that is retarded in phase with respect to the current. This phase shift resembles the action of a capacitance and creates the very large pseudo-capacitance.

The charging of a large pseudo-capacitance for steel electrodes in concrete has to be considered in tests with short duration of the polarisation. It can be difficult to find an acceptable compromise between a polarisation period sufficiently long to reduce this charging current and sufficiently short with respect to time induced changes of the electrodes.

Potentiodynamic scans could not be used to determine the linear polarisation resistance of the cut reinforcement bars in a 15 year old bridge due to the large and unpredictable pseudo-capacitance and the polarisation induced changes of the electrodes. This capacitance caused a charging current that tended to overshadow the Faraday current. It was hardly possible to find a scanning procedure that discriminated between high and low corrosion rates in this case.

Potentiostatic polarisation of steel in concrete leads to a current that decreases with time sometimes to very low values. The polarisation resistance is then dependent on the test time and the current can roughly be expressed by the equation

$$I = I_o \times t^{-0.5} \tag{2}$$

where I is the current, t the time and I_o a 'constant' that varied from test location to test location. The severity of corrosion is related to the 'constant' I_o. It is concluded that the severity of corrosion can be determined from potentiostatic experiments when an empirical base for the evaluations is developed.

References

1. ASTM C876-91 Standard Test Method for Half-Cell Potentials of Uncoated Reinforcing Steel in Concrete.
2. M. Stern and A. L. Geary, *J. Electrochem. Soc.*, 1957, **104**, 56.
3. K. Videm, The reliability of electrochemical techniques for assessing corrosion of steel in concrete. NACE, Houston, Tx, USA. Report 98794, March 1998.
4. K. Videm, Experience with galvanostatic pulse technique and other methods to assess rebar corrosion. *Corrosion '97*, Paper 279, NACE, Houston, Tx, 1997.
5. K. Videm and R. Myrdal, Phenomena that disturb corrosion monitoring of steel in concrete, in *13th Int. Corrosion Congr.*, Melbourne Australia, 1996.

6. K. Videm and R. Myrdal, Instrumentation and condition assessment performed on Gimsøystraumen bridge, in *Int. Conf. on Repair of Concrete Structures*. SND, NPRA, Rescon. Svolvær, Norway, May 1997, p. 375.

7. K. Videm, *6th Scand. Corrosion Congr.*, paper 14, Gothenburg 1971.

8. M. Keddam, H. Takenouti, X.R. Novoa, C. Andrade and C. Alonso, Study of the dielectric characteristics of cement paste *EMCR97, 6th Int. Symp. on Electrochemical Methods of Corrosion Research*. University of Trento, Italy, August 1997.

9. C. Andrade, L. Soler and X. R. Novoa, Advances in electrochemical impedance measurements in reinforced concrete. *Mater. Sci. Forum*, 1995, **192–194**, 843–856.

10. O'M. Bockris and A. K. N. Reddy, *Modern Aspects of Electrochemistry. Vol. 1*. Plenum Press, New York.

11. C. Andrade, Advances in the on-site electrochemical measurement of reinforcement corrosion and their use for predicting residual life (Keynote Address) in *13th Int. Corrosion Congr.*, Melbourne, Australia, 1996.

12. J. A. Gonzalez, A. Molina, L.M. Escudero and C. Andrade. Errors in the electrochemical evaluation of very small corrosion rates — I. Polarisation resistance method applied to corrosion of steel in concrete. *Corros. Sci.*, 1985, **25**, (10), 917–930.

13. S. G. Millard, K. R. Gowers and J. H. Bungey, Galvanostatic pulse techniques: A rapid method of assessing corrosion rates of steel in concrete structures. *Corrosion '95*, Paper 525, NACE, Houston, Tx, 1995.

14. A. A. Sagues. S. C. Kranc and E. I. Moreno, Evaluation of electrochemical impedance with constant phase angle component from the galvanostatic step response of steel in concrete. *Electrochim. Acta*, 1996, **41**, 1239–1243.

Electrochemical Realkalisation and Chloride Removal

11

Repair of Reinforced Concrete Structures by Electrochemical Techniques — Field Experience

B. ELSENER, L. ZIMMERMANN, D. BÜRCHLER and H. BÖHNI

Institute of Materials Chemistry and Corrosion, Swiss Federal Institute of Technology, ETH Hönggerberg, CH-8093 Zürich, Switzerland

ABSTRACT

Electrochemical chloride removal (ECR) and electrochemical realkalisation (ER) are increasingly used as restoration techniques in civil engineering practice. In both cases an alkaline environment is re-established in the concrete pore solution in the vicinity of the rebars. Two well-documented site jobs of successful application — a bridge substructure and a reinforced concrete building facade — are described. The main problems with electrochemical repair techniques on-site, the possible ways to control the effectiveness, the long-term durability and the application to post-tensioned structures are discussed.

1. Introduction

Steel reinforced concrete is one of the most widely used construction materials throughout the world. In general, reinforced concrete (RC) has proved to be successful in terms of both structural performance and durability. This is due to the spontaneous formation of a very thin oxide layer (passive film) on the surface of the steel in the highly alkaline pore solution that protects the steel from corrosion. Chloride ingress from deicing salts or sea water and/or carbonation (the reaction of the alkaline pore water with CO_2 from the atmosphere) lead to depassivation, i.e. the local disruption of the protective passive film on the rebars. In the presence of oxygen and moisture at the steel surface corrosion of the reinforcement starts. Different repair stategies for RC structures with corroding reinforcement can be applied [1]; frequently the chloride contaminated or carbonated concrete around the rebars is removed and replaced. This repair technique ('patch work') has several disadvantages and sometimes is not possible because of static strength requirements. Electrochemical methods such as cathodic protection, electrochemical realkalisation (ER) or electrochemical chloride removal (ECR) involve a d.c. current flow applied to the steel/concrete interface to promote repassivation as a result of hydroxyl ion production and chloride extraction. These repair methods are essentially non-destructive, limit noise to a minimum and only slightly change the concrete surface. Whereas several laboratory studies on realkalisation [2,3] and chloride removal [4,5] exist, only few well-documented site jobs are described in the literature despite the increasing use of these techniques in civil engineering practice [6,7].

In this work, a brief theoretical background of both treatments is given [1,2,5].

Then the results of laboratory research and of two well characterised and documented site jobs — a bridge substructure (ECR) and a reinforced concrete building facade (ER) — are presented. Practical aspects, e.g. the importance of condition assessement before the treatment, techniques available to monitor the effectiveness of the treatment and possible side-effects are pointed out. Finally, the durability of steel repassivation as a means of stopping corrosion, and the application to structures with high-strength steel are discussed.

2. Fundamentals

In all electrochemical restoration techniques a direct current is applied between the reinforcement (cathode) and an external anode in electrolytic contact with the concrete (Fig. 1). Cathodic protection (CP) is a permanent installation with design currents below 10 mAm^{-2}, electrochemical chloride removal (ECR) and electrochemical realkalisation (ER) are applied only on a temporary basis and use currents up to 1 Am^{-2}. In all three cases the *electrochemical reactions at the cathode* (the rebars)

$$2\,H_2O + O_2 + 4e^- \rightarrow 4\,OH^- \tag{1a}$$

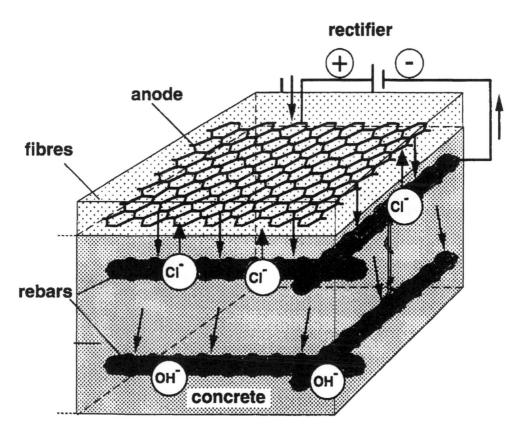

Fig. 1 *Principle of the electrochemical repair methods for steel in concrete.*

$$2 \, H_2O + 2e^- \rightarrow 2 \, OH^- + H_2 \tag{1b}$$

produce hydroxyl ions leading to an increase of the pH near the rebar. This facilitates passivation of the steel. Reaction (1b) is possible at very high current densities and produces hydrogen and especially high tensile steels under load could suffer hydrogen embrittlement.

At the anode the possible oxidation reactions are either chlorine evolution or water decomposition

$$2 \, H_2O \rightarrow O_2 + 4H^+ + 4e^- \tag{2a}$$

$$2 \, Cl^- \rightarrow Cl_2 + 2e^- \tag{2b}$$

$$H_2O + Cl_2 \rightarrow HCl + HClO \tag{2b}$$

These reactions lead to an acidification of the electrolyte around the anode. The decrease in pH value in the electrolyte around the anode depends on the current density applied.

2.1. Ion Migration

The d.c. current flowing from the rebars to the external anode is carried by all cations and anions present in the pore solution of the concrete: negative ions (OH^- and Cl^-) migrate from the reinforcement to the external anode, positive ions (Na^+, K^+) migrate towards the reinforcment. The migration of ions in concrete is a complex process, basically it is governed by the same laws as in aqueous solutions, that is by the electrical field and the fundamental parameters of ion mobility u_i, concentration c_i and charge $|z_i|$ of the ions in the pore solution [8]. According to eqn (3) the different ions in the pore solution contribute in proportion to their concentration c_i and their ionic mobility u_i (Table 1) to the total current. Ion mobilities in concrete are three orders of magnitude lower compared to aqueous solutions; this can be explained by the decrease in available pore volume (aqueous solution 100%; in concrete only about 10%) and the resulting tortuous diffusion path in the pore system [8]. The *chloride transference number*, t_{Cl}, defines the amount of d.c. current carried by the chloride ions, I_{Cl}, in relation to the total current I_{tot}:

$$t_{Cl} = I_{Cl}/I_{tot} = c_{Cl} \, u_{Cl} \, |z_{Cl}| \, / \, \Sigma \, c_i \, u_i \, |z_i| \tag{3}$$

Thus, the chloride transference number t_{Cl} depends on the chloride concentration and on the amount of other ions present in the pore solution, especially OH^- and alkali ions. The composition of pore water solution expressed from chloride contaminated OPC mortar is given in Table 2 [9] from which it is seen that essentially a potassium hydroxide solution is formed. The calculated transference numbers t reported in Table 2 show that most of the current is transported by hydroxides and potassium ions, for the chloride ion $t_{Cl} = 0.10$ is found. Polder [10] reports for mortars with 1% Cl^- by weight of cement a transport number t_{Cl} of *ca.* 0.22, and for lower chloride concentrations 0.12.

Table 1. Ion mobility in acqueous solutions at 25 °C

Ion	Mobility u_i (cm^2 V^{-1} s^{-1})
Cl$^-$	7.91×10^{-4}
K$^+$	7.62×10^{-4}
Na$^+$	5.19×10^{-4}
Li$^+$	4.01×10^{-4}
H$^+$	36.30×10^{-4}
OH$^-$	20.50×10^{-4}
HCO$_3^-$	4.61×10^{-4}

Table 2. Composition of pore water solution for chloride contaminated alkaline mortar [9] with the calculated transference numbers t

Ion	OH$^-$	Na$^+$	K$^+$	Ca^{2+}	Cl$^-$
c mol. L^{-1}	0.5	0.035	0.58	10^{-4}	0.2
t	0.62	0.01	0.27	0	0.10

As a result of the electrochemical reaction at the electrodes and the associated transport processes major changes in the pore water composition will occur during the treatment [5,10–12]. The changes that are beneficial, favouring the repassivation of the rebars, are the increase in OH$^-$ concentration near the cathode (eqn 1), up to 2 mol L^{-1} has been determined [12], and the decrease of the chloride concentration, this being most pronounced at the concrete surface [12]. The increase in pH is higher in presence of chlorides, because a part of the current is transported by the chloride ions [8, 11].

3. Laboratory Results

The feasibility of electrochemical restoration has been shown for both techniques, chloride removal and realkalisation, in laboratory tests. For the *electrochemical chloride removal* all studies show a reduction in the total chloride content between 40 and 60% [4,5,13], the most pronounced decrease in chloride content being found in the cover concrete (Fig. 2). Chlorides behind the first layer of the reinforcement are removed more slowly. As is further shown in Fig. 2, the expected increase in OH$^-$ content (eqn 1) at the rebars actually takes place. Chloride removal is most efficient in the early stages of the treatment (Fig. 3). Prolonging the treatment to more than 1500 Ah m^{-2} is ineffective because nearly all of the current flowing is transported by hydroxyl ions. The increase in OH$^-$ content and the decrease in chloride content result in a marked reduction of the chloride transference number (eqn 3) [4,8,12].

Extensive laboratory tests on *electrochemical realkalisation* have been conducted at BAM Berlin [2,3]. It has been found that the realkalised zone around the rebar increases

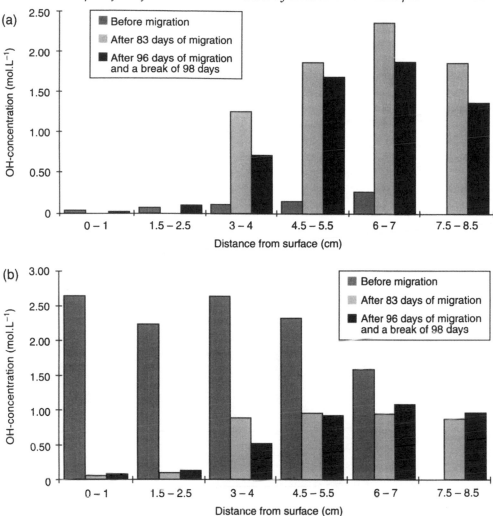

Fig. 2 *Concentration of hydroxides (a) and chlorides (b) in the pore water of cores from a highly chloride contaminated test wall before and after electrochemical chloride removal [5].*

with time and current density (Fig. 4). On the other hand, the observed realkalisation from the concrete surface is independent of current density and thus due only to capillary suction of the alkaline electrolyte (Na_2CO_3) used for the treatment. The total alkalinity of the concrete — determined by acid titration of concrete powder — cannot be increased by electrochemical realkalisation and remains at the low levels of carbonated concrete (Fig. 5). This indicates that electrochemical realkalisation only re-establishes an alkaline pore solution composition; the solid alkalinity reservoir $Ca(OH)_2$ present in uncarbonated concrete cannot be regained.

The corrosion state of the rebars has been tested after the realkalisation treatment and repassivation of the corroding bars has been found after sufficient charge has passed (treatment duration).

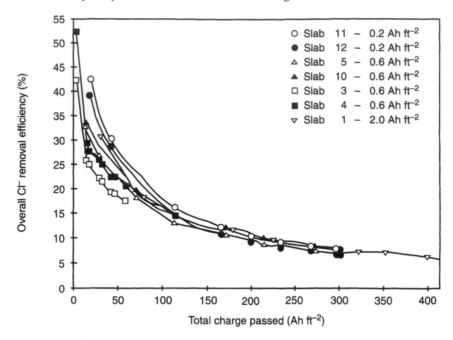

Fig. 3 *Chloride removal efficiency as a function of treatment duration (charge flow) for different current densities [4]. Units: 1 m² = 10.7 ft².*

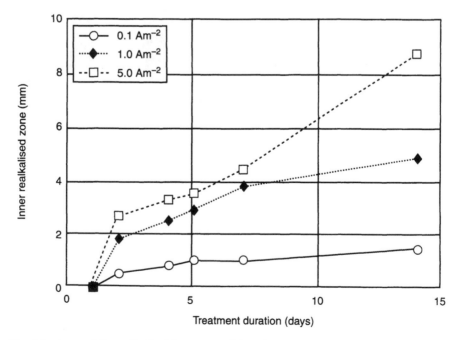

Fig. 4 *Progress of the realkalised front around the rebars, electrolyte Na₂CO₃. Mortar samples w/c ratio 0.6 [2].*

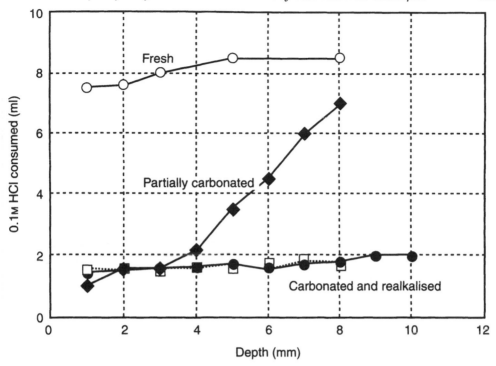

Fig. 5 *Comparison of the alkalinity profiles (consumption of HCl to neutralise concrete powder) for fresh, partially carbonated and realkalised concrete [2].*

4. Results from Field Application

Despite the frequent use of electrochemical restoration techniques in practice only a few site jobs are well documented in the literature. In this Section, two well characterised and documented site jobs managed by the authors — a bridge substructure (ECR) and a reinforced concrete building facade (ER) — are described. The importance of condition assessement prior to the treatment in order to get reliable results on effectiveness and durability is pointed out.

4.1. Condition Assessment

Prior to any rehabilitation of a structure the amount, location and causes of the main damage, i.e. corrosion of the rebars, have to be clearly established. Based on the information from condition assessement and the requirements and future plans of the owner of the structure, a repair strategy can be chosen [1] and the repair project can be planned. Electrochemical repair techniques are feasible when corrosion of the reinforcement is still far from affecting structural safety and only small parts of the concrete surface show damage (e.g. spalling, delamination) that has to be repaired conventionally. Characterisation of the test site and condition assessement were conducted prior to repair with electrochemical methods:

4.1.1. Electrochemical Realkalisation [13]

The reinforced structure chosen for the field experiment with ER was a 5 × 4 m wide part of a large facade of a RC building constructed in 1972. The results of the condition assessements were as follows:

- *Visual inspection* revealed a few areas with concrete spalling and rust staining, in addition to some small delaminated zones. Cracks over the reinforcement were detected close to the sites of corrosion damage.

- *Concrete cover*. An average value of 24 ± 4 mm of concrete cover was found for the vertical reinforcement. The outer horizontal rebars showed an average cover of 15 ± 4 mm. Areas of cover depth < 12 mm corresponded to the areas with concrete spalling and rust formation.

- *Depth of carbonation*. An average of 15 mm carbonation depth was determined from thin section analysis and this agreed well with the observed concrete spalling. Spraying of an indicator solution (phenolphthalein) on drilled powder can overestimate the real carbonation depth [13,14].

- *Potential mapping* was carried out with a $Cu/CuSO_4$ reference electrode (CSE) with a grid size of ca. 25 cm. The average value measured on the visually dry surface was around +50 mV (CSE) but in the vicinity of visible damage values of ca. –200 mV (CSE) were measured. Corroding rebars were indicated on about 15% of the total surface.

4.1.2. Electrochemical Chloride Removal [6,12]

Chloride removal was carried out on one side of a RC subway wall constructed in 1968. The results of the condition assessement were as follows.

- *Visual inspection*. Few small cracks and few areas with insufficient concrete quality.

- *Concrete cover*. An average value of 30 ± 5 mm was measured.

- *Chloride content*. Chloride contents up to 2% by weight of cement at the level of the rebars were found at the bottom of the subway wall. One metre above the traffic lane chloride contents were below 0.6 %.

- *Potential mapping*. An eight-wheel electrode system with a grid size of 15 cm was used. Negative potential values were measured at the bottom of the wall and corresponded to high chloride contents.

On both structures reinforcement corrosion in terms of loss in cross section was negligibly small. The electrical continuity of the reinforcement, assessed by measuring the electrical resistance between two remote points of the area to be treated, was found to be very good and no electrical short-circuits were detected. Areas of spalling, with cracks or visibly bad concrete quality were repaired conventionally with a suitable mortar prior to the electrochemical treatment.

4.2. Electrochemical Treatment

After having installed the electrical connections to the rebars (two on both sites) the anode system was mounted on the concrete surface.

- For the realkalisation job four specially designed plastics tanks about 10 cm thick were mounted with a rubber sealing on the concrete surface. The anode built into the tanks was activated titanium mesh. The electrolyte used was 1M sodium carbonate (Na_2CO_3) solution. A current density of *ca.* 2 Am^{-2} was imposed for 12 days, during which time the resistance of the system (and thus the voltage) decreased due to an increase in conductivity of the concrete pore solution [13,14]. The total charge passed was 1.5×10^6 Cm^{-2} (equivalent to *ca.* 420 Ahm^{-2}).

- For the chloride removal, a titanium anode mesh was mounted on small wooden strips and embedded in a layer of wet cellulose fibres sprayed on the concrete surface. Tap water was used as electrolyte. A current density decreasing from about 0.8 to 0.3 Ahm^{-2} was applied during the treatment time of eight weeks, the total charge passed was *ca.* 5×10^6 Cm^{-2} [12].

In addition to the continuous recording of the rectifier voltage, current density and total charge flow, the contractors drilled small holes in the treated areas of the concrete in order to determine the residual chloride content or the sodium content of the concrete respectively during the treatment.

4.3. Checking the Effectiveness of the Treatment

The goal of electrochemical repair methods is to stop ongoing corrosion of the rebars and to extract chlorides from the concrete cover. At the end of the treatment the effectiveness should be checked. This is usually done by direct measurements (half-cell potential mapping) or by indirect means (charge flow, chloride content, sodium profiles etc.).

4.3.1. Half-cell Potential Measurements

Immediately after switching off the current, potentials below –1.1 V (CSE) were measured and after 1 day the potentials were in the range of –0.7 to –0.9 V (CSE). It was observed that the rebars in the control field (beside the test field) were polarised to more negative values during the treatment period. The process of depolarisation of the steel is quite slow (Fig. 6). After about one month the potentials stabilised around –0.25 V (CSE); at longer times a trend towards slightly more positive potentials can be noted. Thus, meaningful half cell potentials can be measured only several weeks *after* completion of the treatment. In addition, careful interpretation of the results is necessary.

- After *electrochemical chloride removal* the negative potentials indicating corroding areas disappeared, the potential field became more homogeneous.

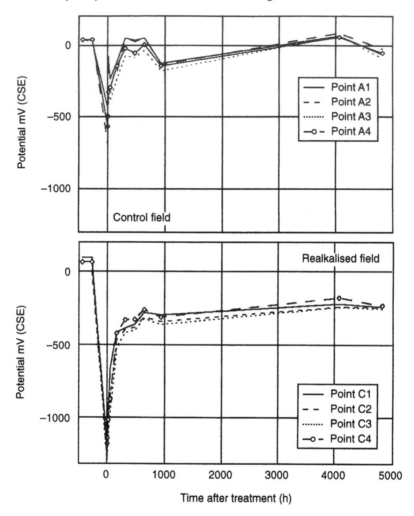

Fig. 6 *Change in half-cell potential measured at the concrete surface in the control and realkalised field [13] of a realkalisation test site at ETH Zurich.*

The average value of the half-cell potentials shifted by about 80 mV to more positive values (Fig. 7), indicating that corroding areas with negative potentials have repassivated.

- After electrochemical realkalisation, half cell potential values of *ca.* –0.2 V (CSE) were measured (Fig. 6), thus the potentials shifted to more negative values by about 250 mV compared to the values of the control field. This behaviour can be understood by taking into account the facts that (a) the concrete cover after the ER treatment has a much lower resistivity (sodium bicarbonate in the pore solution) and (b) passive steel in concrete acts as a pH electrode. The potential values measured indicate that the steel/concrete interface had become more alkaline.

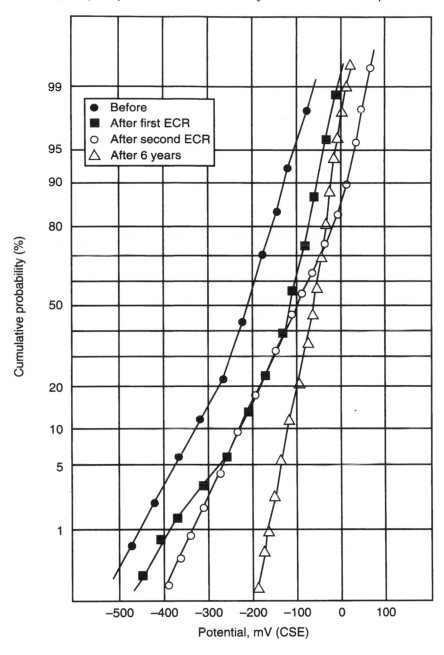

Fig. 7 *Cumulative probability distribution of half-cell potentials measured on the concrete surface of the subway wall prior to, and after electrochemical chloride removal [6].*

The presence of a homogeneous potential field after the treatment is considered a strong indication of repassivation (Fig. 8). In this context it is interesting to note that the potential field over the treated area is already quite homogeneous after 7 days and no signs of the 'corrosion spots' present before the treatment could be observed.

Other electrochemical techniques (galvanostatic pulse, potentiostatic polarisation at a positive potential) have been tentatively used to assess the corrosion state of the rebars.

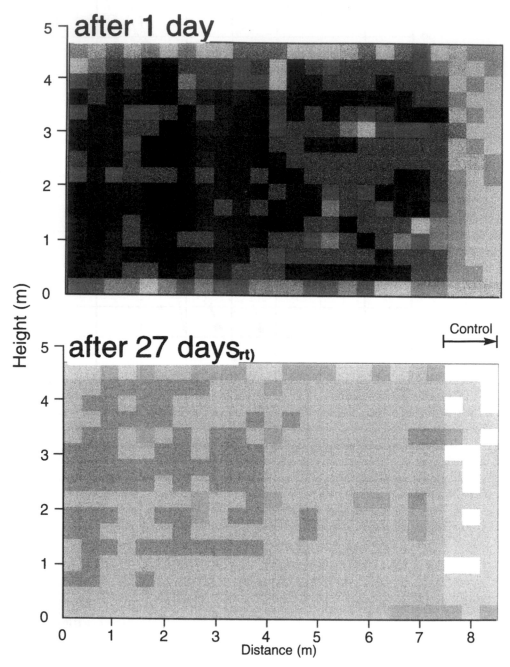

Fig. 8 *Half-cell potential measurements before (top) and one month after realkalisation treatment (bottom). Note that the three lines at the right correspond to the untreated control field.*

(A colour plate of this figure can be found at the end of the book, facing p. 212.)

4.3.2. Total Charge Flow

The mechanistic action of ER and ECR involves the increase in alkalinity at the rebar surface by electrolysis and the migration of ions (chlorides) by electromigration [1,2,5,12]. Both are related to current density or — integrated over time — the charge flow. Laboratory and field tests of ECR have shown that the process of chloride removal becomes inefficient at a total charge flow > 1500 Ahm^{-2} (Fig. 3) and the treatment can then be stopped. It has been suggested by the authors that this might be due to the comparatively slow release of bound chlorides [6,12]. This hypothesis is sustained by the fact that in a second chloride removal treatment 9 months later a further 50% of the total chloride content was removed [12].

In electrochemical realkalisation contractors operate with an empirical value of 450 Ahm^{-2} that must flow during the treatment in order to produce sufficient hydroxyl ions at the rebars. In any case, the total charge flow can only be indicative of an overall quality of the treatment, it gives no local information on reduction in chloride content or repassivation of corroding rebars.

4.3.3. Sodium Content

Chemical equilibrium in the sodium carbonate–bicarbonate system allows one to calculate that a total Na_2CO_3 content > 0.4 mol L^{-1} (corresponding approximately to > 3 kg Na_2O/m^3 concrete) results in a pH of > 10.5 in the solution in contact with CO_2 from the atmosphere. Laboratory experiments have shown that normal steel is passive in such a solution [2]. Thus the determination of the sodium content in depth profiles (Fig. 9) is often used by the contractors as additional criteria for the treatment effectiveness. It has to be noted that the sodium content does not provide direct indication on steel repassivation nor on the pH value around the rebars.

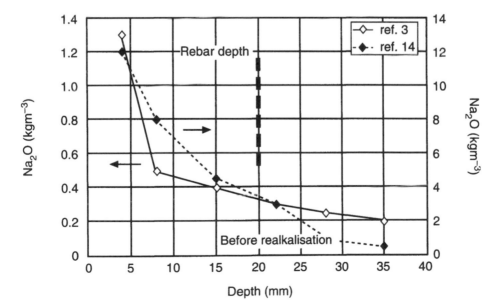

Fig. 9 Sodium profiles taken from cores after realkalisation treatment [3, 14]. Note the marked increase of Na in the cover concrete.

4.3.4. Chloride Content

The total chloride content in the concrete after electrochemical chloride removal can be determined from cores taken after the treatment. On structures with inhomogeneous chloride contamination (frequently the case) it may be difficult to obtain correct information on the amount of chloride removed. It is necessary to analyse several cores and only a statistical treatment of the data can provide the overall efficiency of the treatment [12]. In order to avoid discussions between owner and contractor, number and positions of cores to be taken and the maximum tolerated chloride content after the treatment have to be agreed in the contract.

5. Durability of the Treatment

The durability of electrochemical repairs depends on the maintenance of a stable passive state of the rebars, thus, both a re-carbonation (loss in alkalinity) and an increase in chloride concentration at the steel surface must be avoided. On the two test sites, after six years for the chloride removal and after two years for the realkalisation, the whole treated surfaces were maintained in the passive state.

- After completion of the ECR treatment a thin polymer modified cementitious coating was applied to avoid further chloride penetration. Half-cell potential measurements six years after the treatment (Fig. 7) showed a homogeneous potential field at potentials around –50 mV (CSE). Thus the rebars in the treated area remained fully passive. The same results were found at the Canadian test site (Burlington Skyway) [15].

- After completion of the realkalisation no surface treatment was applied. Half-cell potential measurements after six months, after one year and after two years showed a homogeneous potential field with values around –0.2 V (CSE). Thus, it seems that the presence of the Na_2CO_3 solution in the pore volume of the concrete can maintain a pH value of > 10.5 in contact with CO_2 from the atmosphere, and the rebars will remain passive.

Investigations on other buildings after realkalisation showed that no leaking out of the sodium content in the concrete occurred (except in the outermost 5 mm) over four to six years [15].

6. Application to Post-Tensioned Structures

The high cathodic current densities applied to the steel in concrete result in potentials in the region of hydrogen evolution (Fig. 6). Hydrogen atoms generated at the steel surface either form hydrogen gas molecules or penetrate into the steel. For high strength steel hydrogen embrittlement and brittle failure can then occur without any warning. The risk for hydrogen embrittlement increases strongly above a critical cathodic current density (*ca.* 0.5 Am^{-2}). Whether such high current densities are reached on the high strength steel depends on the current distribution between external anode, the rebar network and the high strength steel, i.e. mainly on the position of the high strength steel and the resistivity of the surrounding concrete. The following points are of importance.

- Electrochemical repair methods should not be applied on structures with prestressed steel reinforcement directly embedded into concrete.

- On post-tensioned structures, where the high strength steel is in a grouted metallic or a plastics duct, different opinions exist on the risk for hydrogen embrittlement during the application of ECR or ER. Further research work is ongoing in order to substantiate the beneficial effect of a metallic duct claimed to act as a Faraday cage and shielding the high strength steel in the duct from the cathodic current. Preliminary results showed a limited current spreadout from the anode. Because of the high risks involved ECR and ER on post-tensioned structures should be applied today only in collaboration with an expert.

7. Conclusions

Electrochemical chloride removal and electrochemical realkalisation lead to an increase in pH at the rebars and to repassivation of corroding steel.

- The durability of ECR has been proven on different site jobs with a track record between five and eight years if further chloride ingress is avoided by applying a coating on the concrete surface.

- Several RC structures treated with ER showed good performance over several years without applying a coating.

- To avoid adverse side-effects the current density during the treatments must be limited to $< 2 \text{ Am}^{-2}$ steel surface.

- Methods and quantitative criteria to assess the efficiency and durability of the electrochemical repair methods should be improved and defined in an international standard.

8. Acknowledgements

Financial support of the Swiss Federal Highway Agency is greatly acknowledged.

References

1. RILEM Draft recommendation for repair strategies for concrete structures damaged by reinforcement corrosion, *Mater. Struct.*, 1994, **27**, 415.

2. J. Mietz, B. Isecke, B. Jonas and F. Zwiener, Elektrochemische Realkalisierung zur Instandsetzung korrosionsgefährdeter Stahlbetonbauteile, Bundesanstalt für Materialprüfung (BAM) Berlin, 1994. Electrochemical realkalisation as repair method for reinforced concrete structures (in German).

3. J. Mietz and B. Isecke, Investigations on electrochemical realkalization for carbonated concrete, *Corrosion '94*, paper 297, NACE, Houston, Tx, 1994.

4. Electrochemical Chloride Removal and Protection of Concrete Bridge Components: Laboratory studies. SHRP S 657, National Research Council, Washington, 1993.

5. J. Tritthart, Elektrochemische Chloridentfernung aus Stahlbeton, Bundesministerium für wirtschaftliche Angelegenheiten, Strassenforschung Heft 459, Wien, 1996. Electrochemical chloride removal from reinforced concrete (in German), and references therein.

6. B. Elsener and H. Böhni, Elektrochemische Instandsetzungsverfahren — Fortschritte und neue Erkenntnisse. "Erhaltung von Brücken", SIA Dokumentation D0129, Schweiz. Ing. und Arch. verein, Zürich (1996). Electrochemical repair methods — progress and new results (in German), pp. 47–59 and references therein.

7. COST 509: Corrosion and Protection of Metals in Contact with Concrete, Final Report, European Commission, Brussels, 1997.

8. B. Elsener and H. Böhni, Ionenmigration und elektische Leitfähigkeit von Beton, Ion migration and electrical resistivity of concrete (in German), SIA Documentation D065, Schweizer Ingenieur und Architektenverein, Zürich, 1990, pp. 51–59.

9. D. Bürchler, B. Elsener and H. Böhni, "Electrical resistivity and dielectric properties of hardened cement paste and mortar", in *Corrosion of Reinforcement in Concrete Construction*, Eds C. L. Page, P. B. Bamforth and J. W. Figg. Royal Society of Chemistry London, (1996), p. 283.

10. R. Polder and H. J. Van der Hondel, "Electrochemical realkalization and chloride removal of concrete", in *Proc. RILEM Conference Rehabilitation of Concrete Structures*, pp. 135–147, Eds D. W. Ho and F. Collins. Melbourne, 1992.

11. L. Bertolini, F. Bolzoni, B. Elsener and P. Pedeferri, "La Rialcalinizzazione e la rimozione dei cloruri nelle costruzioni in cemento armato", *Edilizie*, Nov. 1993 (in Italian).

12. B. Elsener, M. Molina and H. Böhni, *Corros. Sci.* 1993, **35**, 1563.

13. B. Elsener, R. Gabriel, L. Zimmermann, D. Bürchler and H. Böhni, Electrochemical Realkalization — Field Experience, *Proc. Eurocorr '96*, Session 11 OR17-17-1-4. Publ. Soc. Chim. Ind. CEFRACOR, Paris, 1987.

14. R. Gabriel, Elektrochemische Realkalisierung, M.Sc. Thesis, IBWK ETH Zürich, 1996 (in German).

15. A. Roti, Beton-Instandsetzung mittels elektrochemischer Realkalisierung und Chloridentfernung, 2. Int. Kongress zur Bauwerkserhaltung (Berlin), 9.-11.2. 1994. Repair of reinforced concrete with electrochemical realkalisation and chloride removal (in German).

12

The Perception of the ASR Problem with Particular Reference to Electrochemical Treatments of Reinforced Concrete

J. B. MILLER

Norwegian Concrete Technologies, N-0915 Oslo, Norway

ABSTRACT

The status of our knowledge of the alkali silicate reaction (ASR) phenomena is briefly reviewed and the factors important to its development are discussed. Since electrochemical treatments are being used more and more to solve the problems of corroding rebars in concrete, the likelihood of these methods in provoking ASR is examined, and an attempt is made to put the subject into perspective.

1. Introduction

Reactions between aggregates and alkalis are part of the normal hardening process of concrete, and as such are harmless and indeed beneficial. Alkali silica reactions (ASR) are thus to be found to some degree in almost all concrete. ASR can result in both contraction and expansion. For the purposes of the following discussion, however, ASR is understood to mean one or more reactions which occur in concrete between certain types of aggregates and alkali metal ions leading to the production of alkali metal silicate gels, and, under certain conditions, resulting in expansion, and thus cracking, of concrete.

The actual mechanisms involved in the reaction and production of gel, and in the expansion and cracking which may ensue, are not fully understood, and a number of hypotheses are current. The knowledge which exists is largely gained from observation and from parameter testing, and is thus largely empirical in nature.

There is still a lack of sound, proven theory which can be used to properly explain and predict ASR-phenomena. It is nevertheless clear that certain conditions must obtain for ASR to occur. These are:

- the presence of suitable quantities of alkali reactive aggregates,

- the presence of moisture,

- the presence of alkali metal ions,

- temperatures above freezing, and

- sufficiently high pH.

Whether ASR will occur or not, and to what extent, and whether the reaction is destructive or not, will depend on the mutual relationship between each of the conditions. At the present state of knowledge, the following general considerations apply.

1.1. Presence of Alkali Reactive Aggregates

Expansion caused by ASR is a function of the quantity, type and size of aggregate present [1,2]. This is not a simple function, as is shown by the typical 'maps' and expansion curves of Figs 1–3. The figures show that for certain ranges in the content of a given aggregate, expansion is insignificant regardless of the amount of metal ions, or of the amount of reactive minerals in the concrete [1].

Coarse aggregates are also more likely to cause damage than finely divided aggregate. In fact, fine aggregate in the filler size range is often added to reduce or eliminate destructive expansion.

1.2. Presence of Moisture

ASR proceeds within a wide range of moisture contents, though expansion results only within a certain range. In general, for ASR to result in expansion, the relative humidity of the concrete must be higher than 80% and less than 100% [3]. Some observers say that expansion does not occur in saturated concrete (for example, below water). Other sources report damage under apparently saturated conditions. The consensus seems to be that expansion does not occur in truly saturated concrete, though there may be exceptions to this.

What does seem to be common to most cases of destructive ASR, is that the expansion takes place within the concrete mass, rather than in the cover zone. The cracking which is visible at the surface is the result of tensional failure caused by the expansion of the interior. In practice, the effects of ASR in the surface zone proper seem to be confined to half-lens shaped pop-outs caused by local gel formation around the larger reactive particles [2,3]. Probably the reason for this apparent lack of activity in the cover zone is to be found in the prevailing moisture conditions, and in lower pH values.

1.3. Alkali Metal Ions

The ions which take part in the reactions are, in practice, those of sodium and potassium. There is also strong evidence that calcium ions play a part [4,5,18], and that the mutual relationship between the various reacting ions determine whether the reactions will result in expansion or contraction.

Most workers agree that for ASR expansion to occur, a certain critical quantity of alkali metal ions must be present. This quantity seems to be specific to a particular aggregate and to a particular cement. There is thus no universal rule which can be applied as to how much alkali metal ion can safely be allowed. Again, as Figs 1–3 indicate, there are upper and lower limits to both aggregate quantities and amounts of alkali metal ions beyond which expansion becomes insignificant.

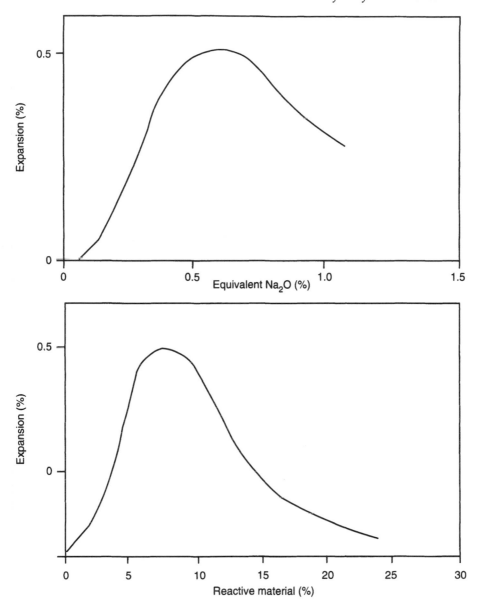

Fig. 1 *General outline of the dependence of mortar bar expansions on the amount of reactive material in the aggregate, given the alkali content, and on the content of equivalent Na$_2$O in the cement, given the amount of reactive material in the aggregate respectively (after [1]).*

1.4. Temperature

ASR does not occur in frozen concrete. In general, ASR otherwise increases in rate with temperature [6], as do most chemical reactions. The order of the reaction does not seem to be clear. Possibly reactions of several orders occur simultaneously.

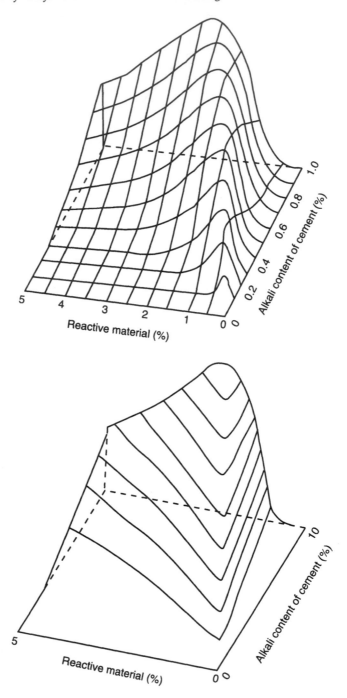

Fig. 2 *A 3-dimensional view of the expansions in relation to active material in the aggregate and alkali content of the cement. The cross-sections for constant reactive material and constant alkali content in the cement respectively are shown at the top. The levels of equal expansion are shown at the bottom (after [1]).*

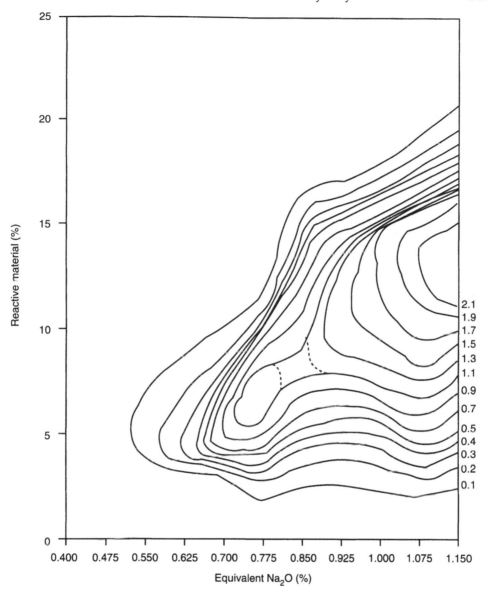

Fig. 3 Iso-expansion map. Each curve shows the combination of reactive material and alkali giving equal expansions. The size of expansions corresponding to each curve is indicated in percentages to the right of the chart. Porous flint. Particle size 1/8 –1/4 mm (after [1]).

1.5. pH Value

It seems clear that ASR only occurs at sufficiently high pH values. It is, for example, well known that ASR seldom, if ever, occurs in carbonated concrete. The required pH value for ASR to occur is not known exactly but from considerations of the known chemistry of the analogous sodium and potassium silicates, it would seem that the necessary pH must be at least 12.5, and is probably higher [7,18].

2. Effect of Desalination on ASR

Desalination, or electrochemical chloride extraction, is regarded by the author as being without influence on ASR, or at most to have only a temporary effect, in actual structures.

During desalination, no alkali metal ions are added to the concrete. However, since the reinforcement is negatively charged, there will be a tendency for alkali metal ions to migrate towards, and concentrate around, the rebars. Since the pH is simultaneously increased, it could be thought that, at some point, ASR could be provoked or accelerated. There exist a number of investigations, chiefly by microscopy, which show, in fact, that this does not occur [8–10]. In some cases, the only effect of the current seems to have been the softening of a few aggregate particles. In other cases, only slight colour changes have been recorded, without there being any important physico-chemical changes.

It may be that the conditions pertaining during desalination are such that the ASR which occurs is of the type which does not result in swelling, i.e. of the type which causes internal cracking by contraction in susceptible aggregates, as opposed to that which produces expansion, and therefore cracking, externally to the aggregates.

It is reasonable to expect, that the increased pH values and alkali metal ion content around the steel, caused by the current, will dissipate with time after the current has been removed. Conditions near the steel will therefore return to normal, save that the chloride ion content will have been drastically reduced, and that the steel itself will have been strongly passivated. The great difference in time scale between the desalination process (weeks) and the effects of ASR (years, decades) should be realised.

The author is aware that it is possible to provoke ASR in the laboratory by applying desalination techniques. However, in the laboratory, conditions are more often representative of the interior masses of concrete rather than of the cover zones of actual structures. Humidities, path lengths, temperatures, and compositions are usually quite different, and are aimed at producing what is known as pessimum conditions. This is, of course, the prerogative of the researcher, but sight of actual structures should not be lost.

ASR may occur at humidities below which expansion does not occur. This sometimes gives rise to a concern that if the structure were to become saturated, as in desalination, the potential expansion would be released in a comparatively short space of time. Under pessimum conditions, this could perhaps happen. In practice, it is unlikely since the time scale of desalination and return to original humidity cycles is short in comparison to ASR phenomena.

Finally, there is considerable evidence that decreasing the amount of chloride ion present in the concrete reduces ASR problems rather than increasing them [18].

3. Effect of Realkalisation on ASR

The pH of carbonated and realkalised concrete is not conducive to ASR, since the equilibrium pH of the 1M sodium carbonate solution introduced is approximately 10.6, which is too low for ASR to occur.

However, if realkalisation is carried too far, there is a danger of sodium being transported beyond the carbonated zone, and into the uncarbonated concrete with its original higher pH still more or less intact.

Under certain conditions, this could be thought to increase the danger or ASR. However, it has been found in practice that transport of alkali metal ions into uncarbonated concrete is difficult and time consuming. Normal realkalisation times of a few days do not seem to allow significant increase of alkali metal ion in the anterior uncarbonated concrete [19]. Nevertheless, to be safe, this transport can be kept under control by pH and sodium monitoring during the realkalisation treatment. Treatment is curtailed when the carbonate solution is shown to have penetrated the carbonated zone, but no further.

Since it is known from studies of realkalised structures that the sodium carbonate introduced into carbonated zones does not migrate away at an easily measurable rate [11,19], the probability of provoking ASR by electro-osmotic realkalisation using sodium carbonate is very slight The reason for the slowness of this migration is probably the comparatively low humidities of cover zones.

4. Other Considerations

On deciding whether to apply electrochemical treatments to concretes where ASR could be provoked the following points should be borne in mind.

1. The reason for applying electrochemical treatments is usually to halt an ongoing corrosion. This being the case, the possibility of provoking ASR tends to become academic. In other words, the real ongoing problem is solved, as opposed to a possible, latent problem which may never occur, and which most likely would have minimal consequences even if it did.

2. Concrete which is already undergoing ASR does not often show corrosion damage as a primary consequence. It is only when ASR has advanced so far that the concrete has been opened up by cracking to salt penetration, carbonation, and freezing and thawing, that reinforcement corrosion becomes a problem. Thus it would seem that electrochemical treatments would still be preferable to allowing corrosion to proceed.

3. ASR seldom leads directly to reduced structural strength. Indeed, one study undertaken by Danish road authorities showed that, for example, shear capacity actually increased as a consequence of ASR [20].

   Problems do arise, however, due to dimensional changes, especially in heavy structures such as dams, and the like, where moving parts may jam or malfunction.

4. With the exception of minor surface pop-outs which can occur under certain conditions, ASR does not usually lead to spalling, unless augmented by secondary agencies, such as re-bar corrosion, or freezing and thawing.

5. Electrochemical treatments are almost always applied to the cover zones of concrete. Any ASR swelling which might be caused in these zones would be slight by virtue of the thinness of the zone compared to the total thickness. As a corollary, it may be said that ASR is of concern only where it occurs in the interior masses of concrete structures. This is not likely to be caused by electrochemical treatment which is applied only to cover zones.

5. Concluding Remarks

1. It has been known for more than 40 years that addition of lithium salts to the concrete mix can be used to control, or eliminate, expansion caused by alkali reactive siliceous aggregates [12,13]. More recently, it has been found that lithium salts can be used on existing concrete to control ASR when introduced from the outside [8,14,15,16,18].

    One way of introducing lithium salts into the concrete is by using electrochemical techniques similar to those used in desalination and realkalisation. Lithium ions may be introduced during desalination or realkalisation by adding judicious quantities to the electrolyte during the treatment. As an aside, lithium may of course be introduced electrochemically as an anti-ASR expansion treatment in its own right.

2. At present, it would seem that desalination has the potential to induce ASR in concrete containing very reactive aggregates [6]. However, this would be the exception rather than the rule. No enhanced ASR activity from actual structures has been reported.

    It is highly improbable that realkalisation can have any pronounced effect on ASR [1,8], and none has been reported from actual structures. The theoretical indications are that it should not, since the pH values involved do not seem to be sufficiently high [1,9], except perhaps fleetingly around the embedded steel.

3. The possibility of inducing ASR when potentially susceptible aggregates are present needs to be weighed against the problem of corroding rebars. Often, solving the problem of ongoing corrosion will weigh more heavily than the risk of inducing ASR. In this connection, it should be borne in mind that desalination, and especially realkalisation, are normally applied to the cover zones of structures, and therefore cannot affect the main bulk of concrete

References

1. P. Bredsdorff, E. Poulsen and H. Spøhr, Experiments on mortar bars prepared with selected Danish aggregates. Committee on alkali reactions in concrete. Progress report I 2, Copenhagen 1966, (DK 666.97.015.83).

2. G. Gudmundsson and H. Asgeirsson, Parameters affecting alkali expansion in Icelandic concrete, in *6th Int. Congr. on Alkalis in Concrete*, Copenhagen, 1983.

3. L. O. Nilsson, Moisture effects on the alkali silicate reaction, in *6th Int. Congr. on Alkalis in Concrete*, Copenhagen, 1983.

4. P. C. Pike and D. Hubbard, Bulletin of the Highway Research Board No. 171, 16 (1959) and No. 175, 39 (1960).

5. S. Chaterji, The role of $Ca(OH)_2$ in the breakdown of Portland cement concrete due to alkali-silica reaction. *Cem. Concr. Res.*, 1979, **9**, London, 1979.

6. Canadian Standard A23.2-14A, March, 1990. Potential Expansivity of Cement-Aggregate Combinations (Concrete Prism Expansion Method).

7. M. Pourbaix, Atlas of Electrochemical Equilibria in Aqueous Solutions. PP 461–463, NACE, 1974, Library of Congress Catalogue Card No. 65-11670.

8. J. Bennett *et al.*, Electrochemical Chloride Removal and Protection of Concrete Bridge Components. Strategic Highway Research Programme, Document SHRP-S-657, National Research Council, Washington, D.C., 1993, p. 113–115.

9. K. C. Natesaiyer, The effects of electric currents on alkali silica reactivity. PhD dissertation, Cornell University, January, 1990.

10. Various investigations by Norwegian Concrete Technologies and licensees — in-house reports, NCT, N-0915, Norway.

11. J. B. Miller, Memo on migration of sodium carbonate from re-alkalised zones. Norwegian Concrete Technologies, Document no. 12/12. 1990.

12. W. J. McCoy and S. G. Caldwell, New Approach to Inhibiting Alkali Aggregate Expansion, *Proc. ACI*, Detroit, USA, 1951, **47**, 693–706.

13. C. S. Forum, Alkali-Aggregate Reaction in Concrete. Christiani & Nielsen, CN Post (published Christiani & Nielsen, Denmark), February 1958.

14. Y. Sakaguchi *et al.*, The inhibiting effect of lithium compounds on alkali-silica reaction, in *Proc. 8th Int. Conf. on Alkali-Aggregate Reaction*, Kyoto, Japan, 1989.

15. C. L. Page, Improvements in and relating to treatments for concrete. UK patent no. BG2 275 265 B. Filed 17th December 1992.

16. J. Bennett, *et al.*, Chloride Removal Implementation Guide. Strategic Highway Research Program, Document SHRP-S-347, National Research Council, Washington, DC, 1993, p. 20–21.

17. P. F. G. Banfill, Realkalisation of carbonated concrete – Effect on concrete properties. *Proc. 6th Int. Conf. on Structural Faults and Repair*, 1995, Vol. 2, p. 277–280.

18. R. Helmuth *et al.*, Alkali-Silica Reactivity: An Overview of Research. Strategic Highway Research Program, Document SHRP-C-342, National Research Council, Washington, D.C. 1993.

19. J. S. Mattila, M. J. Penti and T. A. Raiski, Durability of Electrochemically Realkalised Concrete Structures, *4th Int. Symp. on Corrosion of Reinforcement in Concrete Construction*, Robinson College, Cambridge, UK, 1996.

20. Load carrying capacity of bridges subjected to alkali-silica reactions. Interim report no. 1: The shear strength of concrete beams subjected to alkali-silica reactions. Ministry of Transport, The Bridge Department, Copenhagen, Denmark, July 1986.

13
Experience on Accuracy of Chloride and Sodium Analysis of Hardened Concrete

G. E. NUSTAD

Norwegian Concrete Technologies, N-0915, Oslo, Norway

ABSTRACT

In its work with electrochemical desalination for the rehabilitation of chloride-contaminated concrete, Norwegian Concrete Technologies (NCT) has experienced conflicting results of chloride analyses used to assess the effect of desalination. This paper summarises the results of two independent Round Robin tests conducted to assess the reliability of chloride analysis techniques using concrete samples with known chloride contents as test material. The deviation in results from nominal (true) values indicates a need for better quality control of equipment and analysis procedures. In the light of these findings, questions regarding the accuracy of sodium analysis have been raised and some initial investigations indicate poorer accuracy than anticipated.

1. Introduction

Desalination is an electrochemical repair method used to arrest rebar corrosion in chloride-contaminated concrete by reducing the chloride content of the concrete. The method is also known as chloride removal and as electrochemical chloride extraction. It was introduced commercially by Norwegian Concrete Technologies (NCT) in 1988. The treatment continues until the chloride content has been sufficiently reduced as defined by previously agreed acceptance criteria and determined by chloride analysis of concrete samples.

In the early days of desalination several parties were often involved in chloride analysis of aliquot concrete samples. This often caused conflicting results and since no reference concrete material existed, evaluation became impossible. In addition, the parties involved in the analyses tended to react in a protective manner, hence a constructive discussion and investigation of the results was difficult.

Since correct chloride analyses are of utmost importance in documenting the effect of desalination, a decision was made to produce a series of standardised concrete materials with known chloride contents. The objective was to have this series available for calibration and control of equipment and procedures for chloride analysis.

Other parties have also experienced deviant results of chloride analysis, and in this paper the results from two independent Round Robin tests, in which reference material of known chloride contents were used, are summarised.

2. Nordic Round Robin Test

In cooperation with the Norwegian Building Research Institute (NBI) and Taywood Engineering Ltd. (TEL) in the UK, NCT undertook to produce a series of concrete reference materials with known, documented chloride contents [1,2].

The reference material was then used in a Round Robin test to evaluate three chloride analysis techniques commonly used in Norway for accuracy, precision and reproducibility. This programme was funded by NORTEST, a Nordic body in the field of technical testing and measurements, and five Nordic laboratories participated. The programme was conducted by NBI and the results summarised below are all taken from the Project Report [3].

2.1. Experimental

The chloride analysis techniques selected were the Quantab Test, Chloride Selective Electrode Potentiometry and Volhard Titration.

Each laboratory received three sets of 12 sachets containing concrete reference material, one set for each chloride analysis technique. Within each set, there were four sachets containing the following chloride contents: 0.20, 0.80 and 4.00, all in % by weight of cement. Only the project co-ordinators at NBI knew the chloride content of each individual sachet.

The laboratories were asked to follow the extraction procedure given by the manufacturer prior to the Quantab test, which entails the extraction of fine dust in boiling water. It is simple, quick and renders the Quantab test cheap and amenable to field analyses, where it is often used to judge the total chloride content.

Prior to using Chloride Selective Electrode Potentiometry and Volhard Titration, the laboratories were asked to extract the chlorides by a defined acid extraction procedure which is believed to dissolve all chlorides so that the total chloride content may be determined.

2.2. Results

The results reported from the laboratories (Lab. 1 to Lab. 5) were summarised and discussed for each of the chloride analysis techniques separately. The results are presented in tables showing the mean values of the four parallels, the standard deviations of the means and deviations in % of the nominal values.

2.2.1. Quantab Test

The results are summarised in Table 1. Three laboratories reported unreadable results* with the lowest chloride concentration. The two remaining laboratories reported results revealing very poor accuracy, whilst the standard deviations showed good precision. That means this test is not acceptable for the 0.20% Cl⁻/cement level (which, for example, is important for acceptance of chloride extraction). At the second lowest chloride concentration level, three laboratories reported results which all showed good accuracy and precision, whilst two laboratories reported unreadable results.

---

* For example, as a result of some contamination at the lowest chloride concentration.

Table 1. Results of chloride analyses on concrete reference material using Quantab Test.
* Unreadable results – apparently the result of some form of dominating contamination from an unknown source

Laboratory	Nominal value (wt% of cement)	Mean of four measured value	SD of mean value	Deviation from nominal value
Lab. 1	0.20	_*	_*	_*
	0.80	0.86	0.0076	+8%
	4.00	3.89	0.066	−3%
Lab. 2	0.20	0.68	0.0068	>> 100%
	0.80	0.82	0.0075	+3%
	4.00	3.22	0.014	−20%
Lab. 3	0.20	_*	_*	_*
	0.80	_*	_*	_*
	4.00	2.56	0.18	−36%
Lab. 4	0.20	_*	_*	_*
	0.80	_*	_*	_*
	4.00	4.16	0.02	+4%
Lab. 5	0.20	0.73	0.0025	>> 100%
	0.80	0.75	0.0052	−6%
	4.00	3.34	0.021	−16%

At the highest chloride concentration level, the deviation from the true value varied from +4% to −36%. Except for the results from Lab. 3, all results exhibited reasonable accuracy, precision and reproducibility. Only Labs 1 and 2 had previous experience with the Quantab Test.

All in all, the Quantab Test seems to have varying accuracy, although the precision is good. Since the accuracy improves with increasing chloride content, interference from other ions may explain the poor accuracy at low chloride concentration levels.

2.2.2. Chloride Selective Electrode Potentiometry

The results given in Table 2 show acceptable precision, reproducibility and accuracy with all chloride concentrations, although the standard deviations of the results from Lab. 1 show somewhat poorer precision. The poor accuracy shown by Lab. 5 indicates a dilution error, since about half of the nominal values were reported at all concentration levels. This reflects the care needed when producing calibration curves and preparing test samples. Dilution, pH, and temperature of the test solutions must be the same as those of the calibration solutions in order to avoid errors. The results indicate that satisfactory analyses may be achieved without previous experience since only Lab. 2 had such experience.

Table 2. Results of chloride analyses on concrete reference material using Chloride Selective Electrode Potentiometry

Laboratory	Nominal value (wt% of cement)	Mean of four measured value	SD of mean value	Deviation from nominal value
Lab. 1	0.20	0.17	0.0035	−15%
	0.80	0.66	0.0185	−18%
	4.00	3.28	0.092	−18%
Lab. 2	0.20	0.19	0.0030	-5%
	0.80	0.68	0.0078	−15%
	4.00	3.56	0.013	−11%
Lab. 3	0.20	0.22	0.005	+10%
	0.80	0.80	0.0082	0%
	4.00	3.90	0.00985	−2%
Lab. 4	0.20	0.24	0.000	+20%
	0.80	0.84	0.005	+3%
	4.00	3.82	0.005	−4%
Lab. 5	0.20	0.10	0.0058	−50%
	0.80	0.34	0.0017	−57%
	4.00	1.87	0.0039	−53%

2.2.3. Volhard Titration

The results given in Table 3 indicate that this technique exhibits rather poor accuracy, precision and reproducibility at the lowest concentration level. The results from Labs. 1, 2, 3 and 4 all show good accuracy, precision and reproducibility with the higher concentrations. As all these laboratories had previous experience with this technique, the poorer results from Lab. 5 may be due to this laboratory being without previous experience.

The somewhat unexpectedly poor results with the lowest concentration were traced to a limitation in the Norwegian Standard recommending a titrating solution not sufficiently dilute to allow the determination of low chloride values. A more dilute titrating solution would have improved the accuracy.

3. Dutch Round Robin Test

Conflicting results of chloride analysis have also been experienced by the Civil Engineering Division at the Ministry of Transport, Public Works and Water Management in Holland. The Civil Engineering Division, therefore, had concrete reference material, with both Portland cement and blast furnace slag cement, produced in order to examine the reliability of chloride analysis. Their reference

Table 3. Results of chloride analyses on concrete reference material using Volhard Titration

Laboratory	Nominal value (wt% of cement)	Mean of four measured value	SD of mean value	Deviation from nominal value
Lab. 1	0.20	0.25	0.0094	+25%
	0.80	0.80	0.0014	0%
	4.00	3.78	0.0105	–5%
Lab. 2	0.20	0.22	0.0008	+10%
	0.80	0.79	0.0012	–1%
	4.00	3.66	0.0010	–9%
Lab. 3	0.20	0.32	0.02	+60%
	0.80	0.76	0.0057	–5%
	4.00	3.68	0.0271	–8%
Lab. 4	0.20	0.04	0.0057	–80%
	0.80	0.70	0.005	–12%
	4.00	3.60	0.0141	–10%
Lab. 5	0.20	–0.002	0.0092	>> 100%
	0.80	0.50	0.0019	–38%
	4.00	3.54	0.0238	–12%

material was used in a Round Robin test and the results summarised below are all taken from the paper by Gulikers [4].

A total of 17 Dutch laboratories received nine sachets of reference concrete samples and were asked to conduct chloride analyses in duplicate, according to their standard procedure, and to report the results as chloride content by weight of cement. None of the laboratories knew that the results were to be used in a Round Robin test.

Ten of the laboratories used Volhard Titration, one used Mohr Titration, while Chloride Selective Electrode Potentiometry, Potentiometric Titration and Spectrophotometry were used by two laboratories.

To extract all available chloride, 16 of the laboratories used an acid extraction procedure designed to avoid problems with sulfides from the blast furnace slag which, if not removed, are known to interfere with the determination of chlorides. The remaining laboratory used hot demineralised water in a cyclic elution process. From the results, both methods appeared effective in extracting the total chloride content.

The cement content was determined by analyses at 15 of the laboratories. The remaining two assumed the cement content based upon the average cement content in a normal concrete structure. In general, the reported values for the cement content corresponded with the nominal cement content.

The results summarised in Fig. 1 show the average deviation from the nominal chloride content for each laboratory. Figure 1, which does not distinguish between the various chloride contents, shows that five of the laboratories obtained poor results.

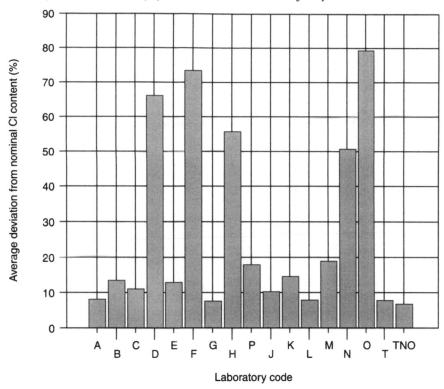

Fig 1. *Results given as average deviation (in %) of nominal chloride content of 18 chloride analyses for the 17 Dutch laboratories [4].*

For one of these, the poor results are related to inaccurate determination of the cement content. For the remaining four, the poor results have no explanation other than doubtful laboratory routines.

According to Fig. 1, laboratory A (Lab. A) seems to have achieved acceptable results. In Fig. 2, the deviation from the nominal chloride contents is given for each of the 18 chloride analyses performed by Lab. A. These show varying accuracy, whilst precision is better. This indicates the great care needed throughout a series of chloride analysis testing.

4. Sodium Analysis

Realkalisation is an electrochemical repair method in which an alkaline solution, normally of sodium carbonate, is transported into carbonated concrete. Although the effect of realkalisation is followed using a pH indicator, additional assessment by sodium analysis is sometimes preferred. Having experienced and documented conflicting results with chloride analysis, the accuracy of sodium analysis was questioned. The accuracy has not fully investigated, but in a test involving four laboratories three of these reported results similar enough to reach the same conclusion, while the results reported from the fourth laboratory led to a completely different conclusion.

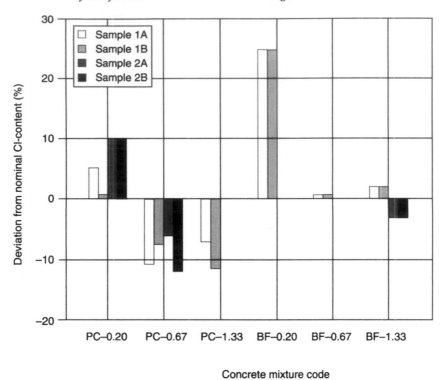

Fig 2. *Deviation (in %) from nominal chloride contents for the 18 chloride analyses carried out by Laboratory A. (PC = Portland cement, BF = Blast furnace cement, 0.20, 0.67 and 1.33 are the chloride contents given as % by weight of cement) [4].*

These results have not been published, since further investigations are required. Also, as no reference material is available for sodium at present, no definite conclusion regarding accuracy is possible.

5. Concluding Remarks

The results from the two Round Robin tests presented confirm the assumption that the accuracy of chloride analysis may vary strongly. The parallel analyses, however, generally revealed very good precision.

In all, the variations in the results call for better laboratory routines for the control of chloride analysis. Use of standardised concrete reference material provides a simple method of improving the reliability of chloride analysis As a consequence of these findings, use of reference concrete is always recommended for desalination projects.

There are indications that sodium analysis may also suffer from poor accuracy, but this has not been fully established. As yet, no reference material for the purpose of sodium analysis has been produced.

Production of standardised concrete material in a properly documented form is time-consuming and demanding. It is therefore recommended that large batches should be produced and the ground material stored in hermetically sealed sachets.

References

1. H. C. Gran and T. Farstad, Production of reference concrete with known chloride contents, Project No. 02877, Norwegian Building Research Institute, Oslo, Norway, 1990.

2. NCT documentation package on the production of standardised concrete dust. (Certificate of Tests on: Concrete Crushing Analysis, Chloride Contents, Moisture Content of Ground Concrete Dust Sample and Analysis of Sample Packets for the Effect on Heating); Norwegian Concrete Technologies, Oslo, Norway, 1992.

3. H. C. Gran, Measurement of chlorides in concrete — an evaluation of three different analysis techniques, NBI Project Report 110-1992, Oslo, Norway, 1992.

4. J. Gulikers, Reliability of chloride analysis in determining the corrosivity of concrete, Workshop US-Europe on Bridge Engineering: Evaluation, management and repair, Barcelona, Spain, 15–17 July 1996.

Part 5

Cathodic Protection

14

Principles of Cathodic Protection and Cathodic Prevention in Atmospherically Exposed Concrete Structures

P. PEDEFERRI

Dipartimento di Chimica Fisica Applicata, Politecnico di Milano, I-20131 Milano, Italy

ABSTRACT

Principles of the cathodic protection of atmospherically exposed concrete structures are described. The various protective effects induced by the cathodic polarisation, the differences between the cathodic protection applied for controlling the corrosion rate of chloride-contaminated constructions and that applied to improve the corrosion resistance of the reinforcement of new structures expected to become contaminated are then underlined and discussed. The more recent applications of cathodic protection to carbonated concrete are also illustrated.

The operating conditions (voltage and current applied), the throwing power, the protection conditions which avoid the risk of hydrogen embrittlement in prestressed structures are also discussed.

1. Introduction

Cathodic protection (CP) of steel reinforcement in atmospherically exposed concrete structures can be achieved by applying a direct current through the concrete from an anode system usually laid on the concrete surface and connected with the positive terminal of a low voltage direct current source to the reinforcement acting as cathode and connected with the negative terminal (Fig. 1). It has proved to be an effective method of stopping or preventing localised corrosion (pitting) of reinforcement caused by the presence of chlorides [1,2].

The first mention of CP applied to structures heavily contaminated with chlorides dates back to the late 1950s [3]. The application of the technique to chloride contaminated concrete started and spread during the 1970s. The first trial to protect bridge decks began in 1973 in North America [4].

Soon, suitable hardware (anodes, overlays, reference electrodes, etc.) had been set up, as well as the protection and design criteria, which turned out to be different from those used in soil and in sea water. In the 1980s, the development of new meshed anodes based, at first, on conductive polymeric materials and then on much more reliable mixed metal oxide activated titanium and also the development of carbon containing paints, led to further applications such as bridge decks, slabs, piles, marine constructions, industrial plants, garages, buildings, etc.

The technique has also been applied to new structures exposed to chlorides, in order to prevent corrosion. This type of CP was named 'cathodic prevention' [5].

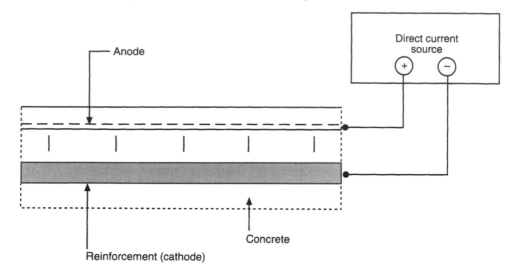

Fig 1. *Schematic representation of the cathodic protection system for concrete structures.*

Even if it uses the same hardware as the traditional CP in concrete, it does experience different operating conditions and side effects. It can be applied to prestressed structures without risk of embrittlement of the high strength steel of tendons. Recently this technique using sacrificial anodes embedded near the periphery of the repair patches has been proposed [2] in conjunction with conventional patch repair of chloride contaminated structures in order to avoid the initiation of incipient pitting around the repaired zones.

Cathodic protection has also proved to be effective in repassivating steel in carbonated concrete [6,7]. It has been applied to carbonated structures containing small amount of chlorides [8].

The cost of CP depends very much on the type of application. In the case of bridge decks the cost has dropped from more than 100 $/m² in the late 1980s to less than 50 in recent years, at least for CP systems which do not need extensive monitoring equipment to ensure they are properly functioning.

Cathodic protection has been applied to more than 600 000 m² of corroding reinforced concrete structures, mainly in North America, where CP has been installed on more than 500 bridges, but also in Europe, in the Middle East and Australia) and to about 150 000 m² of new and to nearly all prestressed ones (mainly in Italy in the period 1990–93).

Although CP gives the option of rehabilitating rather than replacing the damaged structures allowing huge cost savings, and cathodic prevention can be a cost effective alternative to other supplementary preventive methods, if evaluated on the basis of the life cycle costs, the wider adoption of the techniques has not been as big as expected. The reason is mainly associated with their high initial cost and the need for continuous functioning and with permanent maintenance and monitoring programmes.

This paper highlights some results achieved in the field of the CP of steel reinforcements embedded in concrete through the research activities and field applications assistance carried out at the Department of Applied Chemical Physics of Politecnico di Milano [1,5–19]. The paper illustrates only the principles and operating conditions of CP. Other contributions of the Department research on anodic and on monitoring systems not discussed here are reported in a bibliography [20–23].

2. Beneficial Effects

The principal beneficial effect arises from the shift of the potential of reinforcing steel in the negative direction. The potential to which CP has to bring the steel in chloride-containing concrete to reach protection conditions is often erroneously considered to be the same as in soil or seawater. This is because CP in concrete is not applied to maintain the steel in immunity conditions, as in the case of traditional CP, but to restore passivity conditions or at least to favour the build-up of passivation phenomena if the steel is already corroding, and to maintain it for the entire service life of the structure.

Other beneficial effects are associated with the oxygen reduction on the steel surface and to the current circulation in the concrete: the former increases the OH^- concentration, the latter reduces the chloride content in the concrete near the steel. While the effects associated with the lowering of potential cease immediately if the current is interrupted, the changes of composition at the surface produced by the cathodic reactions or by the migration of ionic species inside the concrete can last for months. This fact gives rise to the possibility of applying CP intermittently or "of applying very high currents for periods of several days or weeks with a view to achieving more persistent protection of the sort associated with electrochemical realkalisation or electrochemical chloride extraction" [7].

3. Behaviour of Steel in Chloride-Contaminated Concrete

Before considering separately the case of the initiation of corrosion and its prevention and that of its propagation and its control, it is important to describe the behaviour of steel in chloride-contaminated concrete.

Different domains of potentials and chloride contents can be identified which correspond to different behaviour of steel (Fig. 2). The extent of these domains depends also on the OH^- content near to the steel surface, temperature, cement type and content, concrete porosity, etc. Consequently the diagram of Fig. 2 has to be considered as only schematic and indicative. The domain A (corrosion zone) indicates the conditions which cause initiation and stable propagation of pits; the domain B ('imperfect' passivity zone according to Pourbaix's definition) the conditions which do not allow the initiation of new pits but the propagation of pre-existing ones; domain C ('perfect' passivity zone) the conditions which do not allow either the initiation or the propagation of pits; the domain D (hydrogen evolution zone) indicates the conditions where hydrogen evolution and consequently hydrogen embrittlement of high strength steel can take place.

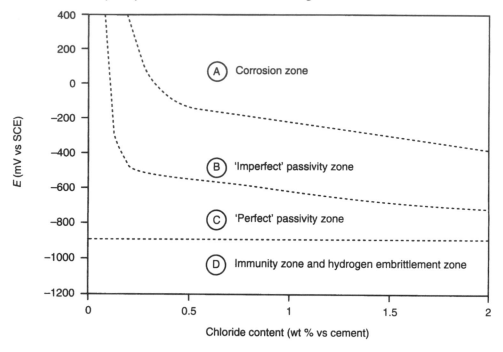

Fig 2. *Schematic illustration of steel behaviour in concrete for different potentials and chloride contents.*

4. The Initiation of Corrosion and its Prevention

When the concrete chloride content increases, the interval of potentials in which steel is passive contracts. The highest potential of this range, called the pitting potential (E_{pit}), typically falls from +500 to –400 mV (vs SCE) on passing from non contaminated to very heavily chloride contaminated concrete (Fig. 2). The highest chloride content compatible with passive conditions for each potential is the critical chloride content at that potential.

For the usual corrosion potential (around 0 vs SCE) the critical content is in the range 0.4–1% of cement weight. By lowering the potential, the critical chloride content increases, as shown in Fig. 2, typically by almost one order of magnitude for a 100 mV decrease with respect to the free corrosion potential.

This strong influence of the potential on the critical chloride content is the basis of the cathodic prevention of pitting of steel reinforcement of structures which are likely to become chloride contaminated. Thus, by imposing a small cathodic polarisation to the rebar of new structures or to structures not yet corroding and maintaining it throughout their entire service life, the critical chloride content is increased with respect to non-polarised structures, so that in practice it will never be reached.

Usually cathodic prevention operates with current densities in the range 1–2 mAm^{-2} which produce a decrease of the potential of about 100–200 mV and consequently an increase of the critical chloride content by more than one order of magnitude.

In Fig. 3 a typical development path (① → ② → ③) (in terms of potential and

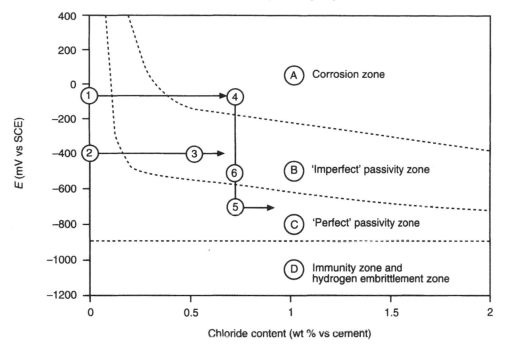

Fig 3. *Development paths of potential and chloride content on the rebar surface of an aerial construction during its service life for: cathodic prevention (①→②→③); CP restoring passivity (④→⑤→); CP reducing corrosion rate (④→⑥→). Cathodic prevention is applied from the beginning, CP only after corrosion has initiated.*

chloride content) of cathodic prevention is shown. At the usual current densities in the range 1–2 mAm^{-2}, a decrease of the potential of at least 100–200 mV is produced leading to an increase of the critical chloride content by more than one order of magnitude with respect to non-polarised structures, so that in practice it will never be reached when applied from the outset over the entire life.

Figure 4 shows the results of application of cathodic prevention to the slabs subjected to a ponding with a NaCl solution. After about 700 days when the chloride content at the steel surface had reached more than about 1%, the initiation of rebar corrosion was detected on the control slab (in the free corrosion condition). On the other hand, a very low current density of 0.4 mAm^{-2} on the reinforcement was enough to lower the potential by about 150–200 mV so that after more than 3 years the initiation of corrosion has not yet taken place although the chloride content on the steel surface has reached values higher than 2% with respect to cement weight. Obviously, the higher the applied cathodic current density, the lower the steel potential and the higher the chloride threshold to initiate corrosion.

5. The Propagation of Corrosion and its Control

If, for a given chloride content, the rebar potential is more positive than E_{pit} or, for a given potential, the chloride content is higher than the critical value, initiation and

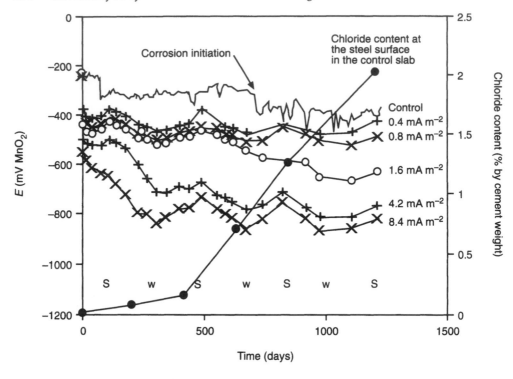

Fig 4. Instant off potential of non-corroding steel cathodically polarised with different current densities and chloride content in the control slab. Specimens exposed to NaCl alternating ponding (3% NaCl in the first 2 years and a saturated solution thereafter), consisting of wetting for one-week followed by drying for two weeks. Control slab started corroding after about 700 days (S = summer, W = winter).

propagation of pits will take place. Once the attack is initiated, it can propagate even at potentials more negative than E_{pit}. To stop corrosion, it is necessary to reach a lower potential (E_{pro}) below which steel repassivates (Fig. 3). The value of E_{pro} , as does E_{pit}, varies with: chloride content, pH on the corroding surface, temperature, etc., remaining in any case about 300 mV lower than E_{pit}.

In the potential interval (E_{pit} – E_{pro}) the propagation of existing pits can occur, although the initiation of new pits is prevented. Nevertheless, a CP which keeps corroding steel in this range of potentials can be of benefit because it reduces the driving voltage between cathodic and anodic areas as well as the extent of the cathodic regions around the corroding regions and thus also the current exchanged between them. This leads to a reduction of the rate of penetration of pits and the restoration of passivation phenomena which eventually, when the potential reaches E_{pro} leads to passivity conditions on the steel. In conclusion, if the potential is maintained in the range (E_{pit} – E_{pro}), especially if it approaches E_{pro} the corrosion rate of existing pits can become low enough to make it acceptable and in any case the initiation of new pits will be prevented.

To follow the variation in the potential of steel reinforcement of a concrete structure damaged by chlorides and then protected by a CP system it is helpful to refer to the map of Fig. 3. The initial condition is represented by the symbol ① where the chloride content is nil and the steel is passive. By increasing the chloride content, the working point shifts to symbol ④ within the corrosion region. Corrosion of the steel occurs rapidly by a macrocell mechanism. The CP leads to ⑤ so that the passivity is restored or to ⑥ without restoring passivity. In all cases the corrosion rate is reduced.

In the case of atmospherically exposed constructions current densities in the range 5–15 mAm^{-2} are generally needed to stop corrosion or to make its rate acceptably low.

Much lower current densities are required in conditions where the oxygen transport to the embedded steels is restricted as in water-saturated concrete (for instance for components operating underwater) so that the oxygen diffusion limiting currents are very low (typically in the range 0.2–2 mAm^{-2} of reinforcing steel area for immersed structures). In these cases the application of currents higher than the limiting values, though still very low, causes hydrogen evolution and consequently bring the potential to very negative values.

The experience on bridge decks shows that the current required to maintain protection conditions (verified by the so called 4 h 100 mV potential decay empirical criterion that is usually used [24]) decreases even after months or years from start-up. This happens because the cathodic current can bring about repassivation of steel active zones, by improving the ratio [OH$^-$]/[Cl$^-$] which increases E_{pro} and/or also because the reduction of the macrocell current exchanged by the inhibition of the macrocell itself [19].

In the cases in which the CP path runs according to ④ → ⑤ → of Fig. 3 and thus passivity conditions are reached on the entire surface of the steel, the current required to maintain passivity is reduced to a few mAm^{-2} (e.g. 2–5 mAm^{-2}). If the CP path runs according to ④ → ⑥ → and thus passivity conditions are reached only on steel surfaces close to the anodes (e.g. only on the surface of the first row) the decrease of current is smaller and the current density to fulfill the protection criteria remains high [19].

6. Cathodic Protection in Carbonated Concrete

The carbonation process reduces the pH of concrete from values higher than 13 to values lower than 9. Consequently, steel passes from passive to active conditions, and thus can corrode provided oxygen and water are available. Theoretically, to stop corrosion of active steel, i.e. the condition in which the metal has no tendency to pass to the oxidised form, potentials more negative than –850 mV vs Cu/CuSO$_4$ should be reached. At the beginning of the 1990s it was shown that in carbonated concrete current densities in the usual CP range which cause a modest lowering of steel potential, can produce enough alkalinity to return the pH at the reinforcement surface from values lower than 9 to values higher than 12 (Fig. 5) and thus from corroding to passive conditions [6].

Recent trials [7] have shown that in carbonated concrete with or without small quantities of chlorides, passive conditions can be reached in five months with current densities of *ca.* 10 mAm^{-2}. Measurements carried out using pH sensitive probes show that the pH changes are confined to the concrete not more than 1 or 2 mm from the steel surface. A current density of 5 mAm^{-2} has produced initial signs of passivation

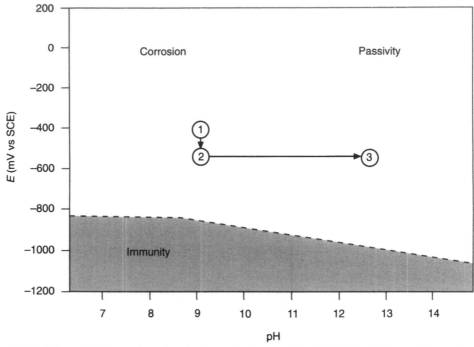

Fig 5. *Schematic illustration of evolution paths of potential and pH when CP is applied to rebar in contact with carbonated concrete [6].*

after only about one year, while a current density of 2 mAm^{-2} has not done so (Fig. 6). Once passivity conditions are reached the current densities to maintain them are lower.

7. Hydrogen Embrittlement

High strength steels used in prestressed constructions (but not the ordinary steel for reinforced concrete) can be subjected to hydrogen embrittlement if their potential is brought to values at which hydrogen evolution can take place. However, with the exception of chromium-containing steel, the risk of hydrogen embrittlement of prestressing steels is low, provided the potentials are more positive than –900 mV [25, 26]. Obviously, the problem does not exist if the high strength steel is shielded by ducts, at least providing the ducts are not interrupted as happens in some types of construction.

Cathodic protection applied to corroding steel works mainly in the region ⑤ of Fig. 3, which is very close to the hydrogen evolution zone, and this can make problematic the application of the technique to complex prestressed structures. On the contrary, CP applied to non-corroding steel works mainly in the ①→ ②→③ route; which is far from the hydrogen evolution and this is one of the reasons this technique can be safely applied to prestressed structures.

The throwing power plays an important role, as discussed in the next section, in the effective operation of CP.

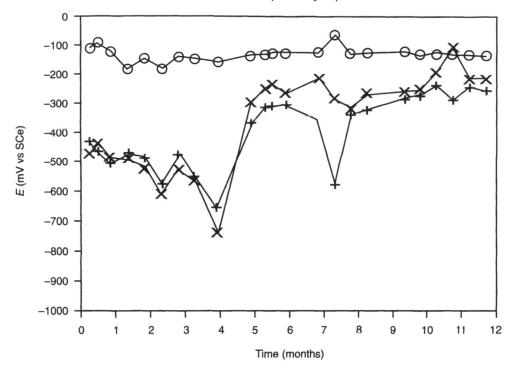

Fig 6. *Potential values of steel in alkaline (o) and carbonated concrete without chloride (×) and with 0.4% chloride by cement weight (+), protected with 10 mAm^{-2}, measured four hours after switching the current off, as a function of time and type of concrete [7].*

8. The Throwing Power

In CP of corroding reinforcement, a uniform current is difficult to achieve because of the high electric resistivity of concrete, the small distance between anode and reinforcement, the low polarisability of corroding steel and the complex geometries of reinforcements. The use of distributed anodes contributes to mitigating the problem for the first reinforcing mat, although the distribution on the deeper ones remains uneven. For this reason, the inner active reinforcement may not be able to be passivated.

The throwing power depends mainly on the steel condition, the effect of resistivity of the concrete being less important. With passive steel, the current distribution and even more the potential distribution improve because the polarisability of passive surfaces, especially at low current densities, is much higher than that of corroding areas.

To select the proper CP working conditions without producing the risk of overprotection on the first mat and to operate safely, the range of potential has to be considered. The lowest potential to be adopted is −900 mV for high strength and −1100 mV (SCE) for ordinary steels [26]. Taking this into consideration, different protection conditions will be achieved, depending on the throwing power. In the presence of corroding steel, the protection condition is obtained (i.e. in zone C or in the lower part of zone B of Fig. 5) without overprotection on steel of the first mat, on reinforcement 25 cm deep on ordinary steel and 15 cm deep in the case of high strength steel.

Similarly, in the presence of non-corroding steel, cathodic prevention (i.e. to operate in the zone B of Fig. 5), is achieved on reinforcement 80 cm deep for ordinary steel and 60 cm deep in the case of high strength steel. The higher throwing power and the less negative potentials that are needed for cathodic prevention allow one to operate with avoidance of the risk of hydrogen embrittlement also on constructions of complex geometry.

In any case, in the presence of high strength steel, CP should be monitored not only to avoid underprotection conditions but also to avoid overprotection [27].

9. Conclusions

Cathodic protection has proved to be an effective method to control chloride -induced corrosion of reinforced concrete structures exposed to the atmosphere, even in the presence of high chloride levels. The risk of hydrogen embrittlement makes its application to prestressed structures possible only in cases of simple geometry and under strict monitoring control (unless high strength steel is shielded by continuous ducts).

Cathodic prevention has proved to be a viable and safe technique to increase the corrosion resistance of reinforcement in reinforced or prestressed structures.

Cathodic protection in carbonated concrete can take the pH on the rebar surface from values lower than 9 to values higher than 12 and thus transforms the state of the reinforcement from corroding to passive conditions.

References

1. P. Pedeferri, *Construction and Building Materials*, 1996, **10**, 391.
2. C. L. Page, Cathodic Protection of Reinforced Concrete — Principles and Applications, in *Proc. Int. Conf. on Repair of Concrete Structures*, Svolvaer, Norway, 28–30, 1997, pp. 123–132.
3. R. F. Stratfull, *Corrosion*, 1957, **13**, 173t.
4. R. F. Stratfull, *Mater. Perform.*, 1974, **13**, 24.
5. P. Pedeferri, Cathodic Protection of New Concrete Constructions, in *Proc. Int. Conf. on Structural Improvement through Corrosion Protection of Reinfored Concrete*, Institute of Corrosion, Leighton Buzzard, UK, 1992.
6. P. Pedeferri, *L'Edilizia*, 1993, **10**, 69.
7. L. Bertolini, F. Bolzoni, T. Pastore and P. Pedeferri, Cathodic Protection of Carbonated Concrete Structures, *Proc. Int. Conf. on Understanding Corrosion Mechanism in Concrete*, MIT, Boston, 1997.
8. G.Areddia, L. Bertolini, L. Lazzari and P. Pedeferri, submitted to *L'Edilizia*.
9. P. Pedeferri, Cathodic Protection and Cathodic Prevention of Aerial Concrete Structures, in *Proc. Int. Conf. on Corrosion in Natural and Industrial Environments: Problems and Solutions*, NACE Italian Section Grado, 1995, pp.45–57.
10. P. Pedeferri, Cathodic Protection of Post-tentioned and Prestressed Structure, *Electrochemische Schutzverfahren für Stahlbetobauweke*, SIA, ETH, Zurich, 1990.

11. T. Pastore, P. Pedeferri, L. Bertolini and F. Bolzoni, Current distribution problems in the cathodic protection of reinforced concrete structures, in *Rehabilitation of Concrete Structures*, Eds D. W. S. Ho and F. Collins, RILEM/CCSIRO/ACRA, Melbourne, pp.189–200, 1992.

12. L. Bertolini, F. Bolzoni, A. Cigada, T. Pastore and P. Pedeferri, *Corros. Sci.*, 1993, **35**, 1633.

13. P. Pedeferri, Principles of cathodic protection and its application to steel in concrete, *Proc. Int. Workshop on Corrosion and Protection of Metals in Contact with Concrete*, COST 509, Orta San Giulio (Italy), 7–8 June 94, Ma51-57, 1994.

14. L. Bertolini, F. Bolzoni, T. Pastore and P. Pedeferri, Investigation on reinforced slabs exposed to chloride solution ponding under cathodic polarization, in *Proc. Int. Conf. on Corrosion in Natural and Industrial Environments: Problems and Solutions*, NACE Italian Section, Grado, 1995, pp.291–300.

15. A. Bazzoni, B. Bazzoni, L. Lazzari, L. Bertolini and P. Pedeferri, Field application of cathodic prevention on reinforced concrete structures, *Corrosion '96*, Paper 312, NACE, Houston, Tx, 1996.

16. L. Bertolini, P. Pedeferri, T. Pastore, B. Bazzoni and L. Lazzari, *Corrosion*, 1996, **52**, 552.

17. L. Bertolini, F. Bolzoni, T. Pastore and P. Pedeferri, New experiences in cathodic prevention of reinforced concrete structures, in *Corrosion of Reinforcement in Concrete*, C. L. Page *et al.*, Society of Chemical Industry, London, 1996, pp.389–398.

18. L. Bertolini, F. Bolzoni, T. Pastore and P. Pedeferri, Three year tests on cathodic prevention of reinforced concrete structures, *Corrosion '97*, Paper 244, NACE, Houston, Tx, 1997.

19. L. Bertolini, F. Bolzoni, T. Pastore and P. Pedeferri, presented to *J. Appl. Electrochem.*

20. C. J. Mudd, P. Pedeferri and G. L. Mussinelli, M. Tettamanti, *Mater. Perform.*, 1988, **27**, 18.

21. P. Pedeferri, G. L. Mussinelli and M. Tettamanti, Experience in Anode Monitoring System For Cathodic Protection of Steel in Concrete, in *Corrosion of Reinforcement in Concrete*, Eds C. L. Page, K. W. J. Treadaway and P. B. Bamforth, pp. 498–506, Society of Chemical Industry, London, 1990.

22. T. Pastore, P. Pedeferri, G. L. Mussinelli and M. Tettamanti, New developements in anode materials and monitoring system for cathodic protection of steel in concrete, *Proc. 11 Int. Corrosion Congr.*, Florence, Vol. 2, pp.467–472, 1990.

23. T. Pastore, P. Pedeferri, M. Tettamanti and J. T. Reding, Potential measurements of steel in concrete using membrane probes, *Corrosion '91*, Paper 119, NACE, Houston, Tx, 1991.

24. European Draft Standard, EN 12696-1, "Cathodic Protection of Steel in Concrete. Part I: Atmospherically Exposed Concrete".

25. C. L. Page, Interfacial effect on electrochemical protection methods applied to steel in chloride containing concrete, in *Rehabilitation of Concrete Structures*, Eds D. W. S. Ho and F. Collins, RILEM/CCSIRO/ACRA, Melbourne, pp.179–187, 1992.

26. S. Klisowski and W.H. Hartt, Qualification of cathodic protection for corrosion control of pretensioned tendons in concrete, in *Corrosion of Reinforcement in Concrete Construction*, Eds C. L. Page, P. B. Bamforth and J. W. Figg, pp.354–368. The Royal Society of Chemistry, Cambridge, 1996.

27. B. Bazzoni and L. Lazzari, A new approach for automatic control and monitoring of cathodically protected reinforced structures, *Mater. Perform.*, 1992, **31**, 13.

15
Cathodic Protection of Reinforced Concrete Structures in The Netherlands — Experience and Developments

R. B. POLDER

TNO Building and Construction Research, PO Box 49, NL-2600 AA Delft, The Netherlands

ABSTRACT

An overview is given of the application of cathodic protection to reinforced concrete structures in The Netherlands. Three cases of successful application of cathodic protection (CP) to concrete structures are described. One example (a large number of cantilever beams with mixed in chloride) shows the relative stability of the performance of a CP system with activated titanium strip anodes in cement-injected boreholes over 6 years. The second case with conductive coating on slabs and columns (with mixed in chloride) was successful for several years and a similar design has been adopted for more structures of the same type. In the third case, mild steel reinforcement was corroding in parts of a post-tensioned structure due to de-icing salt leakage in the joint. The reinforcement was cathodically protected with a one metre wide strip of conductive coating anode, without endangering the prestressing steel. During design, execution and monitoring of these and other projects, testing methods and evaluation procedures have been developed, which were adopted in a national Technical Recommendation. In all documented cases, CP has shown to be effective in preventing further corrosion damage. Long term monitoring of structures receiving CP has shown that significant scatter may be present in sets of depolarisation values. Drying out of the concrete (due to repairs and the application of a coating) causes the current to decrease, sometimes also reducing the depolarisation levels. Both effects pose difficulties for interpretation, which require further study.

1. Introduction

Cathodic protection (CP) of reinforcing steel in concrete structures has been used successfully for over 20 years. Its history and the various stages of its development have been described by Pedeferri [1]. Cathodic protection is able to stop corrosion in a reliable and economical way where chloride contamination had caused reinforcement corrosion and subsequent damage to the concrete. To new structures where corrosion induced damage is anticipated, cathodic prevention is applied [2]. Recently the state-of-the-art was described [3] and a draft European standard has been published [4]. In The Netherlands, CP was introduced in 1987 with two small trial projects. Since then, one or two full scale projects were made every year until

1993. In all cases, alternatives such as replacement of precast concrete elements or conventional repair were considered. Cathodic protection was preferred for reasons of practicability, low impact on the users of the structure, safety and durability. In 1994 three, in 1995 six and in 1996 five installations were completed and energised. This growth was due to more economical materials (such as conductive coatings) and increased awareness of the advantages of CP. Based on the experience gained in most of the Dutch projects and additional laboratory research, a CUR Technical Recommendation was published [5]. In 1996, the Ministry of Transport (Rijkswaterstaat) has applied CP to parts of a post-tensioned bridge. By the end of 1996, the total number of installations in The Netherlands was about 20. In contrast with most other countries, most CP cases concern mixed in chloride and relatively small precast concrete elements. This paper describes three examples and collects some of the observations made during the design, execution and monitoring of the installations.

2. Cantilever Beams in Two Apartment Blocks, Tilburg

2.1. History

This case concerns 2448 precast concrete cantilever beams supporting balcony slabs in two housing blocks of 17 floors at Tilburg, The Netherlands, built in 1972. Many beams had suffered severe corrosion damage due to calcium chloride added to the fresh concrete as a set accelerator. The average chloride content (acid-soluble) was 0.86% by weight of cement, with considerable variation. The cement was ordinary Portland cement. Cover depth to the steel was about 30 mm. Cracking of concrete occurred after less than 10 years service. A polyurethane coating was applied after polymer impregnation, but in a few years damage reappeared. In 1990, it was realised that a better solution was needed. Replacement of the elements would have required heavy work on the facade (timber window frames, brick masonry). It was considered that this would cause too much inconvenience to the residents, so it was decided to apply cathodic protection.

2.2. Design and Execution

The anode layout was unconventional, with a strip of activated titanium in a hole drilled along the main axis of the beams and filled with cementitious grout, as shown in Fig. 1. As an alternative anode design, titanium mesh on the surface embedded in a cementitious overlay was considered. However, a mesh/overlay design appeared to be more expensive and might meet problems due to the impregnation applied when the concrete was coated. Furthermore, an overlay would cause changes of profile of the beams, which was judged unacceptable for the residents. The drilled-in anode was tested before the proposed design and was finally accepted. In laboratory tests of current distribution to individual rebars in similar beams, the optimal anode layout was found [6,7]. As the rebar cages had been welded, the steel continuity in all elements was good and the cover to the steel showed little variation. Continuity between elements was not always present, so a reinforcement connection

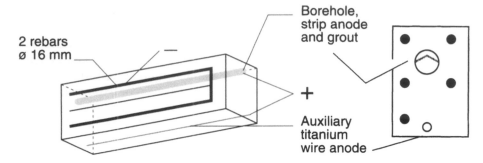

Fig. 1 *Cantilever beam with strip anode in cement-filled borehole, Tilburg.*

was necessary in each beam. The system would operate under constant current control. At the time when the design was made, it was considered necessary to control the current accurately and on a detailed level. Moreover, the commercially available rectifiers had limited capacity. Based on these considerations, 68 zones of 18 beams in each building were made, separated for practical reasons in East and West wing and for orientation North and South.

A hollow coring drill of 35 mm diameter was used for making the boreholes. To account for the possibility that the anode would touch a reinforcing bar (short circuit) or be close to one, which were both regarded undesirable, a multi-step testing scheme was applied.

(i)   The absence of short circuits was tested using a d.c. resistance device when the anode strip was placed in the borehole and again immediately after injection with grout. If a short circuit was detected (less than 1 Ω), the anode was removed and the steel in the borehole was located. Around that spot, the anode was isolated with a piece of plastics pipe.

(ii) After some days hardening, polarisation was applied for a short time to each beam and the subsequent change of steel potentials was measured with respect to a reference electrode on various positions on the concrete surface. The potential shift values were similar on all locations on the surface, showing that the (short term) current distribution within the beams was quite uniform. Some cases where connections between rebars and negative wiring was inadequate were easily detected by this test and corrected.

(iii) The a.c.-resistance of all beams (anode to cathode) was tested before energising. The variation for all beams, disregarding the effect of different lengths, was between 20 and 100 Ω. If different beam lengths were accounted for, all resistances lay between 0.5 and 2 times the mean value, suggesting that they all had been executed well.

(iv) After commissioning, a few short-circuits could be detected because the driving voltage of the particular zone (necessary to achieve the pre-set current level) was significantly lower than the normal value. These shorts had been caused by incorrect wiring, which was traced and corrected easily.

2.3. Conventional Repairs

Before CP application, significant cracking and spalling were present and mortar repairs (of various size) were necessary in about 25% of the total number of beams. To meet the need for uniform current distribution, such patches should have similar current conduction (or resistivity) to that of the original concrete. Several repair mortars were checked for resistivity. One mortar had a very high resistance and was rejected; the others were accepted. After slight modification (see below), the testing procedure was used in later projects and laid down in [5].

2.4. Performance During Six Years

Depolarisation readings with respect to manganese dioxide (MnO$_2$) reference electrodes (RE) have been taken at least twice a year since energising the two systems at the end of 1990. Figure 2 shows average data of 34 zones of one installation. In a few zones minor adjustments to the current levels were made to reduce depolarisation values that were too high or to increase relatively low depolarisation. The average current density is about 8 mA per m^2 steel surface, corresponding to 35 mAm^{-2} anode surface, which is well below the maximum anode current density level (108 mAm^{-2}). The driving voltages show some variation with the season, from 1.3 V in summer to 2.5 V in winter. This is due to the higher resistivity of concrete at lower temperatures [8,9].

For one building, the average depolarisation values were between 110 and 150 mV (after 4 h) and between 180 and 250 mV (after 24 h); they show a trend of slow increase over the years of about +5 mV per year, with considerable fluctuations. On two occasions, depolarisation measurements were carried on for about 9 days. These 'long term' depolarisation values were substantially higher (+15 to 25%) than those after 24 h. This shows that evaluating the depolarisation of the steel over a 24 h

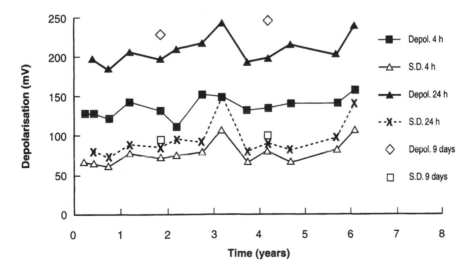

Fig. 2 *Average (depol, solid lines) and standard deviation (S.D., hatched lines) of depolarisation for 4 or 24 h (9 days) of 34 zones, each consisting of 18 precast cantilever beams, Tilburg.*

period is on the safe side. Within each set of depolarisation data of 34 zones at one point in time, considerable variation was present. Typical standard deviations were 80 mV for an average of 140 mV. Probably this is due to local differences in wetting (normal exposure to rain, some leakage of joints or drains). All winter datasets have higher standard deviations than the preceding or following summer datasets. This was found in both buildings, so the cause must be an external effect, probably connected with the weather.

The results show that CP in both buildings has performed satisfactorily. The high stability of the electrical results in the course of time may be caused by the relative absence of environmental influences. The anode is embedded deep inside the concrete and the elements are sheltered from most of the rainfall by the slabs on top. The temperature is probably the most widely varying factor. In another CP project, with the anodes in an overlay on the upper side of balconies exposed to rain, much stronger seasonal variations with season were observed [10]. The reference electrodes seemed to perform well as none was found to fail. A few of the rectifiers failed, probably as a result of a lightning strike.

3. Gallery Slabs and Frames, Groningen

3.1. History and Preliminary Investigation

This apartment flat was built in 1957. It consists of gallery slabs and rectangular supporting frames, which were precast on-site, on one side of the structure. The concrete had been coated since the 1960s. In the early 1980s, cracking due to corrosion appeared, which was repaired with polymer mortar. Cracking and spalling reappeared in about 5 years. Cover depths were 20–30 mm and carbonation was 0–20 mm. The total (acid-soluble) chloride content was between 0.3 and 0.8% (average 0.6%) by mass of cement, apparently added to the concrete mix. The cement was ordinary Portland cement. Steel continuity was poor, probably because of corrosion of binding wires. Corrosion was still active and the damage was expected to increase. Serious conventional repair and replacement of the elements were considered. Both alternatives would be very costly and would strongly disturb the residents with dust, noise and limited access to their apartments for a considerable time. Consequently, CP was chosen as the most favourable option.

3.2. Design and Execution

Because the exposure to rain of all the units is similar, one electrical zone of 600 m² concrete surface was made. As drawings were absent, the steel surface area was unknown. It was considerably less than 1 m² of steel surface per m² of concrete surface. Old polymer repairs were removed. Spalls were repaired using a trowelling mortar or a flowing mortar (both cementitious with polymer modification). Steel continuity was achieved by cutting grooves in the concrete to the steel and welding steel wires to all bars. A conductive primer with a semi-conductive polymer and a relatively low graphite content together with twisted silver wire ribbon as a primary anode made up the anode system. A cosmetic top coat was applied. The layout is given in Fig. 3. A normal polymer floor coating was applied to the top surface of the slabs.

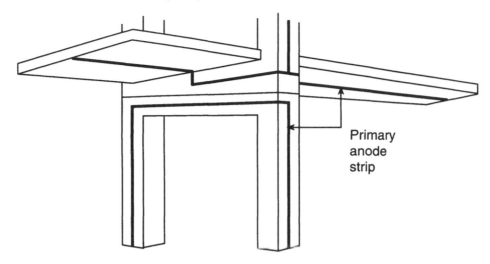

Fig. 3 *Layout of slabs and frames with conductive coating on all faces except top face of slabs, Groningen.*

Fourteen graphite REs were evenly distributed over the slabs and the frames (and over the height of the building). The system operates under constant voltage control, with a maximum of 2.0 V. Potentials were monitored by datalogger with remote control and data acquisition by telephone.

3.3. Repair Mortar Compatibility

As mentioned before, repair mortar should not prevent uniform distribution of the protection current. Mortars should be 'electrically compatible', meaning that they should have similar electrical resistivity. Resistivity data for repair mortars are generally not available. For practical resistivity testing, the test should not have a very long duration. Specimens should be small enough to exchange moisture quickly with the environment. The number of specimens should account for some mortar mix inhomogeneity. A more complicated specimen used in concrete resistivity tests [11], which was based on pioneering work by others [12], was simplified for routine mortar resistivity testing. Specimens were $100 \times 100 \times 50$ mm^3 prisms with two embedded brass bars. For each of the two proposed mortars, four specimens were exposed in a fog room (20°C and >95% relative humidity (R.H.)) and four in a climate room with 20°C and 80% R.H. The resistance between the brass bars was measured with 108 Hz a.c. for eight weeks. Readings were converted to resistivity using cell constants obtained with solutions of known conductivity. The exposure climates represent two extreme situations: very wet concrete in an exposed structure (fog room) and a sheltered structure in equilibrium with the annual-average R.H. in Western Europe (20°C 80% R.H.).

This procedure has been adopted in the Dutch Technical Recommendation [5], including reference values for concrete resistivity as a function of exposure and cement type, (see Table 1). These reference values were based on long term laboratory tests in various exposure climates [11,13] and field experience [14], all with various cement

Table 1. *Reference values according to [3, 5] for resistivity of mature concrete as a function of cement type, environmental exposure and properties or surface treatment of concrete; corresponding laboratory climates in { }*

	Concrete resistivity (Ωm)	
Cement type	Ordinary Portland Cement (CEM I)	Blast furnace slag cement (>65% slag, CEM III/B), fly ash (>20%), silica fume (>5%)
Very wet, submerged splash zone {fog room}	50–200	300–1000
Outside, exposed	100–400	500–2000
Outside, sheltered; *coated, hydrophobised, not carbonated* {20°C. 80% R.H.}	200–500	1000–4000
Ditto, *carbonated*	1000 and higher	2000–6000 and higher
Indoor climate *carbonated* {20°C. 50% R.H.}	3000 and higher	4000–10 000 and higher

types. The definition of similar resistivity is that the mortar resistivity after 8 weeks in the fog room or 20°C/98% R.H. is not more than twice and not less than half the mean value of the concrete, either as measured on site or from cores, or according to the appropriate reference value in this Table.

For comparison, two cores were taken from the structure and the a.c.-resistance was measured between steel plates pressed to the two faces via cloth which was lightly impregnated with soap solution, after exposure for some time in 80% R.H. and later in the fog room. The resistivity was calculated by simple geometric conversion. The cores had a resistivity of 600–1200 Ωm in 80% R.H. and about 250 Ωm in the fog room. Both dry and wet values are relatively high for Portland cement concrete [3,11], which is probably due to partial carbonation. For judging the resistivity of the repair mortars, the value in 80% R.H. was considered most important because of the sheltered exposure of the structure. In 80% R.H., the resistivity of the flowing mortar was about 650 Ωm and that of the trowelling mortar about 1200 Ωm. Both mortars were judged to be electrically compatible.

3.4. Performance

The system was commissioned in September 1994 and the electrical operation was evaluated intensively during the first four months. The average depolarisation in

24 h was 150 mV, which is satisfactory with respect to the criterion (>100 mV). The standard deviation ($n = 14$) was 72 mV. The current density was relatively low: 600 mA flow through 600 m^2 concrete surface (containing possibly 300 m^2 steel surface). Considering the presence of the coating system and the moderate rain exposure, this was regarded as reasonable. It was concluded that the system operates satisfactory. Later measurements (3 to 4 per year) up to three years after commissioning have confirmed this. From 1993 to 1996, CP has been applied according to the same design to four buildings of this type with similar results.

4. Abutments of a Post-Tensioned Bridge, river Dommel

4.1. History

In 1996, the Dutch Ministry of Transport (Rijkswaterstaat) applied CP for the first time to a concrete structure. This involved two parallel post-tensioned bridges of 14 m wide, suffering from corrosion resulting from de-icing salt leakage in the abutment joints. Chloride had penetrated deeply into the underside of the bridge deck over about a width of half a metre from the joint (Fig. 4). Conventional repair by cutting out spalls and all contaminated concrete and replacing with new concrete was considered to be technically possible. However, the chloride contaminated concrete would have had to be cut out behind the reinforcement bars. This would have impaired the structural capacity and required temporary support, which would have involved lifting of the bridge deck. Such work would have serious impact on the traffic, and was considered unacceptable. Furthermore, the working space was limited to less than half a metre height, so the required cleaning of the steel and application of shotcrete could probably not have been applied with the necessary attention. Consequently, the new concrete had to be cast from the top of the deck. Again, this

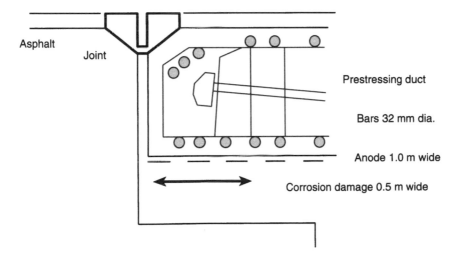

Fig. 4 *Cross-section of a bridge abutment, river Dommel.*

would disturb the traffic. In fact, the cost of regulating traffic to ensure safety for both passing vehicles and personnel involved in the repair process tends to become substantial. If CP were to be applied, the chloride contaminated concrete could be left in place. Only spalls would need to be removed and made good with a mortar whose function would be only to provide electrolytic conduction. Considering all the possibilities, CP was preferred as the least disruptive method with good expectation of durability. If the cost of traffic measures are included, CP was certainly more economic.

4.2. Design and Execution

The structure was made using OPC. The surface area of the reinforcement in the zone to be protected was approximately equal to the concrete surface. Cover depth was about 30 mm and carbonation depth was practically zero. The presence of prestressing steel was taken into account during the design of the system as follows. In the abutments, the post-tensioning steel was lying in ducts at a depth from the underside of at least 250 mm. It was anticipated that the current density at the ducts would be low, so strong polarisation was not to be expected [15]. It was decided to monitor the polarisation of the ducts by embedding reference electrodes in their vicinity. Consequently, the CP system design was as follows. A conductive coating anode was applied to a zone of one metre wide from the joint and a silver wire ribbon primary anode was installed parallel to the joint. The total length comprised four separate parts of 1 m wide by 14 m long (2 North, 2 South). As the exposure was expected to be quite homogeneous, the four parts were put together in one electrical zone. Sixteen REs for monitoring the polarisation of the post-tensioning steel were installed at the depth of the ducts. For expected optimal long term performance and stability, manganese dioxide REs were chosen. Close to the mild steel reinforcement, sixteen graphite REs were placed for normal protection monitoring. The monitoring frequency is four times per year. The criteria were:

- protection of reinforcing steel: (average) depolarisation in 24 h > 100 mV; and

- safety of prestressing steel with regard to hydrogen embrittlement potentials: (all individual) polarised potentials more positive than –850 mV vs Ag/AgCl; considering the scatter in the base potentials of the REs (generally ±10–20 mV around the mean), it was decided to set the safety limit 50 mV more positive than the limit of –900 mV vs Ag/AgCl according to [3].

4.3. Performance

The system was energised in October 1996. The performance during five months is summarised as follows. The average depolarisation of reinforcing steel over 24 h has been > 100 mV since about 3 months from the start, at relatively mild driving voltages of about 1.8 V, despite low temperatures (about 0°C). In March 1997 (+8°C) the average depolarisation was 120 mV; the standard deviation was 37 mV ($n = 16$). The depolarisation over 4 h is less in all cases (average 80 mV in March 1997). Apparently this system needs more than a few hours to depolarise; judging depolarisation for short periods like 4 h is not a good criterion here and will lead to overprotection. The polarisation of the prestressing ducts is very mild; typical values are about −200 mV vs Ag/AgCl. The current distribution was relatively stable over time. One field draws significantly more current than the other three (+25%). This may have been caused by persistent local leakage in the joint. After the first five months, the system was considered to operate satisfactorily.

4.4. Further Results

In June 1997, the depolarisation of the mild steel was below 100 mV (average about 90 mV). In order to improve the level of depolarisation, the driving voltage was deliberately increased to 2.5 and for a short time to 3.0 V. The depolarisation improved. However, in August 1997, during a long hot and dry period with temperatures near 30°C, the current density was very low (less than 1 mA m^{-2}), clearly due to increased concrete resistivity; the average depolarisation was well below 50 mV. In October the average depolarisation was again over 100 mV (at 2.5 V with a current density about 3 mA m^{-2}). In all cases, significant scatter was present within the measurements. After one year of operation, despite short term deviations from the pre-set criterion (100 mV depolarisation), the system was judged to operate satisfactorily. It is expected that over the years the depolarisation will improve systematically.

5. Discussion of Evaluation Procedures

The scatter in depolarisation datasets as found in all three examples and the variation in particular in dry and warm periods, raise the question of evaluating practical CP systems. It is clear that during hot and dry periods, the protection current decreases if a system is operated under constant voltage. At the same time, the resistivity increases, reducing the corrosivity of the steel environment. In particular in locations exposed to high temperatures and strong drying out, it does not seem sensible to allow the driving voltage to be increased too much. So in situations where drying out is significant, systems working under constant current control will also find difficulty in operating. Secondly, the 100 mV depolarisation criterion is empirical. There is no scientific reason why a structure showing say 90 mV depolarisation would develop (certain) corrosion damage, while the same structure with 100 mV depolarisation, would not (with the same certainty). Finally, the scatter in depolarisation datasets (at one point in time) pose the problem of having a satisfactory average result, with individual readings below the required value. Disregarding individual observations would not really solve this problem. It is felt that these matters require further study by the international engineering and scientific community.

6. Concluding Remarks

In The Netherlands cathodic protection has been applied successfully to about 20 structures during the past ten years. Cathodic protection was shown to stop corrosion of reinforcement effectively, including cases where previous conventional repairs had not stopped further development of corrosion induced damage. In the medium to long term, CP may save considerable money compared to (repeated) conventional repairs. The design and execution require specific expertise and attention. Repair materials must have similar electrical resistivity to the concrete of the structure. Test methods and reference data are available. Various anode types are available and different layouts have been developed to suit the particular requirements of many structures and their owners, showing that CP is a flexible method. Over the last ten years, the design of installations could be simplified because of better understanding and increased confidence in the technique. The number of zones could be reduced (or the size of zones enlarged) due to improved insight in the role of the electrical resistivity of concrete. It appears to be safe to evaluate depolarisation over 24 h instead of 4 h. The experience was laid down in a national Technical Recommendation, giving guidelines for many practical issues. In a recent case it was found that the presence of prestressing steel in ducts does not have to preclude the use of CP. In this case, the polarisation of the prestressing ducts was shown to have a comfortable safety margin with respect to hydrogen embrittlement potentials, while the reinforcing steel is sufficiently protected.

It is recognised that considerable variation may be present in depolarisation datasets. Drying out of the concrete as a result of the repairs undertaken and of the application of a coating may have a serious effect on the resistivity of the concrete, on the current density and on the level of depolarisation subsequently found. However, with drying out the corrosivity decreases. It is not clear from the existing documents, how such a situation should be treated. The procedures for evaluation of the quality of the protection need improvement, both from a corrosion technology and a statistical point of view.

7. Acknowledgement

The owners of the described structures, their advisors and the contractors involved are gratefully acknowledged for their permission to publish the information.

References

1. P. Pedeferri, Concrete structures: Cathodic protection and cathodic prevention, *Eurocorr'97*, Trondheim. This volume, p. 161–171.
2. L. Bertolini, F. Bolzoni, T. Pastore and P. Pedeferri, New experiences on cathodic prevention of reinforced concrete structures, in *Proc. 4th Int. Symp. on Corrosion of Reinforcement in Concrete Construction*, Eds C. L. Page, P.B.Bamforth, J. W. Figg, Society of Chemical Industry, 1996, 389–398.
3. COST 509, Corrosion and protection of metals in contact with concrete; final report, Eds R. N. Cox, R. Cigna, Ø. Vennesland and T. Valente, European Commission, Directorate-General Science, Research and Development, EUR 17608 EN, 146 pp., 1997.
4. CEN, Cathodic protection of steel in concrete — Part 1: Atmospherically exposed concrete, prEN 12696-1, December 1996.
5. CUR, Kathodische bescherming van wapening in betonconstructies, (Cathodic protection

of reinforcement in concrete structures), CUR Technical Recommendation 45, Gouda, in Dutch, 1996.

6. P. C. Nuiten and R. B. Polder, Current distribution in multi-element cathodic protection systems, Construction Maintenance and Repair, July/August, 8–13, 1991.

7. R. B. Polder and P. C. Nuiten, A multi-element approach for cathodic protection of reinforced concrete, *Mater. Perform.*, 11–14, **33**, 1994.

8. D. Bürchler, Der elektrische Widerstand von zementösen Werkstoffen, Ph.D. Thesis nr. 11876, ETH Zürich, 142 pp, 1996.

9. L. Bertolini and R. B. Polder, Concrete resistivity and reinforcement corrosion rate as a function of temperature and humidity of the environment, TNO report 97-BT-R0574, 1997, 85 pp.

10. P. C. Nuiten, Successful cathodic protection of 288 Dutch balcony elements, *Construction Maintenance and Repair*, 1990, 178–181.

11. R. B. Polder and M. B. G. Ketelaars, Electrical resistance of blast furnace slag and ordinary Portland cement concretes, in *Proc. Int. Conf. on Blended Cements in Construction*, Ed. R. N. Swamy, Elsevier, 1991, pp 401–415.

12. J. Tritthart and H. Geymayer, 1985, Aenderungen des elektrischen Widerstandes in austrocknendem Beton, Zement und Beton, **1**, 74–79.

13. R. B. Polder, 1996, Laboratory testing of five concrete types for durability in a marine environment, in *Proc. 4th Int. Symp. on Corrosion of Reinforcement in Concrete Construction*, Eds C. L. Page, P. B. Bamforth, J. W. Figg, Society of Chemical Industry, London 1996, pp.115–126.

14. R. B. Polder, P. B.Bamforth, M. Basheer, J. Chapman-Andrews, R. Cigna, M. I. Jafar, A. Mazzoni, E. Nolan and H. Wojtas, Reinforcement Corrosion and Concrete Resistivity — state of the art, laboratory and field results, in *Proc. Int. Conf. Corrosion and Corrosion Protection of Steel in Concrete*, Ed. R. N. Swamy, Sheffield Academic Press, 1994, pp.571–580.

15. P. Pedeferri, Personal communication, 1996.

16
Ten Years of Cathodic Protection in Concrete in Switzerland

CH. HALDEMANN and A. SCHREYER

Helbling Ingenieurunternehmung AG, Hohlstr. 610, CH-8048 Zurich, Switzerland

ABSTRACT

Since the 19th century buried structures have been successfully cathodically protected against corrosion. For approximately the last 10 years, concrete bridges and tunnels are also increasingly being repaired by the use of cathodic protection to combat the corrosion resulting from their exposure to chloride penetration. The method has proved to be particularly successful in bridge pillars and tunnel walls. Results in the construction area in Switzerland are summarised and a survey given of the use of preventive corrosion protection in other countries.

1. Introduction

Construction projects such as bridges, tunnels and buildings require large financial commitment and represent an immense capital investment for a country. One of the fundamental materials used in these infrastructures is reinforced concrete. The long life span of this construction material is primarily due to the fact that steel in the concrete is protected by a thin oxide layer which develops in the high pH conditions of the pore water in cement. The penetration of chlorides, which contact the concrete structures through the use of road de-icing salt, as well as the advancing carbonation of concrete lead to a local destruction of this oxide layer. If oxygen and water also reach these places at the same time, the steel starts to corrode. Such corrosion intrusion affects the middle and long-term safety and usage of the structures. The classical method of restoring such damage is costly and time consuming. Since the chloride infected concrete must be removed (usually down to the first steel layer) the strongly corroded iron has to be replaced and then again refilled with concrete.

Because of static strength or traffic reasons such type of restoration is often impossible to achieve in practice. A renovation without having to remove large amounts of old concrete is often needed. Cathodic protection against corrosion (CP) has been proved as a superior and appropriate method of stopping corrosion processes even in concrete containing large amounts of chlorides. This paper surveys the last 10 years results of the use of CP in concrete building structures in Switzerland and elsewhere with some selected examples from Switzerland.

Cathodic protection has been used for decades as a protection measure and has been proved successful in many branches of industry. Ships have been protected in this way for more than 100 years against corrosion. Cathodic protection with impressed current has been used since the beginning of the 20th century in the protection of ground and water pipelines. It has been known since the 1950s that concrete can also be used as an electrolyte and that CP is therefore appropriate in concrete construction. Stratfull, in 1973, after a number of tests successfully protected a bridge plate with CP in the USA [1]. In 1985 the first CP protected bridge with impressed current in Switzerland was installed by the Helbling Ingenieurunternehmung AG. Since then, more than a million square metres have been protected by CP throughout the world.

2. Cathodic Protection Procedure and Criteria of Protection

Cathodic protection is an active protection procedure in which the electrochemical corrosion cycle is electrically influenced. The method uses the fact that the rate of corrosion is dependent on the electrochemical potential. Through the impact of an adequately high protective current from a d.c. source (usually a rectifier) the potential of the protected part of the structure is shifted in the negative direction thus reducing the rate of, or preventing, the corrosion. Inert-anodes are used to inject the protective current into the structure, the negative pole of the rectifier being connected to the protected metal and the positive pole to the anode. The literature has several detailed descriptions of how CP functions [2] and, in addition, national and international norms and guides have been set [3] or are in preparation [4].

Inert anodes serve to introduce the protective current delivered by the rectifier. In Switzerland anodes of titanium in nets or strips are being used for this purpose.

The required level of protective current depends on external factors such as condition of the rebars, chloride level, concrete humidity etc. For maximum protection to be achieved the protection area may be divided into various zones which are separately controlled.

3. Survey

The actual guidelines for CP recommend permanent monitoring. This provides the possibility of checking the functioning of the system and of noting any changes in the factors, such as concrete humidity and chloride level that can influence the corrosion process. Such changes can then be discovered in good time for the protective current to be suitably adjusted. The following monitoring elements are installed for checking the effectiveness of CP.

- **Reference electrodes.** In concrete buildings the installation of silver/silver chloride electrodes is approved. These serve to measure the electrochemical potential of the rebars (on and off switching potential) and the depolarisation performance.

- **Macrocells.** For additional control of the efficiency of the CP in several installations so-called macrocells are installed. These are pieces of steel which are implanted in high-level chloride mortar, cemented in the building and connected through copper wires to the rebars. Through the external wiring it is possible to measure the current density and direction and to control the functioning of CP. In addition probes that are in danger of severe corrosion can be removed after a certain time and their condition checked.

- **Temperature and humidity sensors.** The building temperature and the concrete humidity determine the electrical conductivity of the concrete; the knowledge of these factors permits better interpretation of data from the reference and measuring sensors.

It is advisable to reduce the number of embedded probes to a minimum and to optimise the frequency of measuring. The monitoring concept is of utmost importance; the probes must be installed in an area which would give a meaningful reading. The newest generation of rectifiers permits an automatic operation of the control measurements and an automatic adaptation of the protective current to the required values. The survey expenses are reduced to a minimum. Such rectifiers can be recommended for large projects although for small surface areas the installation cost may be too high.

4. Checking the Efficiency

An absolute potential value as a protection criterion has proved unreliable. In Switzerland both the following protection criteria are relevant and are advised in Rule C7 of the Swiss Company for Corrosion Protection (SGK) [3]:

- **Depolarisation.** For the so-called 100 mV criterion the difference between the potential immediately after turning off the protective current (U_{off}) and after being shut off for 4 h ($U_{off,4h}$) is the measurement to be made. As soon as this difference is greater than 100 mV this criterion has been fulfilled.

- **Measurement sensors.** As soon as the measurement probe indicates the presence of the protective current, the CP-system will be functioning at this point. If current flow is also indicated in the other relevant areas (high chloride levels, high humidity, etc.) it can be accepted that the whole structure is receiving enough cathodic protection.

American tests have shown [6] that even when the protection criteria cannot totally be achieved the corrosion can be permanently stopped.

5. Experiences

5.1. Foreign Countries

The USA and Canada show world-wide the largest and longest results with CP protected concrete structures. Recently in a report of the Strategic Highway Research Program (SHRP-S-337) [7], it was noted that in the USA more than 350 structures, mostly bridges, with a total surface of approximately 940 000 m^2 are cathodically protected and 90% of the installations function totally satisfactorily. The remaining 10% are showing defects due to owner or installer's mistakes, for example, fuses being defective over an extended period, cut wires, etc. In 1982 the Federal Highway Administration had already officially confirmed that CP is the only restoration technique which proves successful in stopping the corrosion in chloride-contaminated bridges regardless of the level of chloride [8].

In Europe, Italy is the leading nation in the use of CP to protect concrete structures, with more than 100 000 m^2 cathodically protected. Contrary to the practice in Switzerland, CP in Italy is primarily used as a preventive measure for new structures. Bridges, especially in the alpine freeways (e.g. Torino Frejus), which are directly affected by road de-icing salt are preventively protected. This deals with the whole road surface as with New Jersey profiles. A titanium net is laid on the completed concrete structure and covered with a 2.5 cm layer of hard concrete (Fig. 1). Because of the use of CP the usual seal for the street plates was no longer necessary. For the preventive CP for the New-Jersey profiles titanium strips were attached with plastics fasteners to the rebars and the concrete then poured in one operation. Cathodic protection increases the service life of the structure.

With the use of suitably controlled rectifiers and accompanying survey monitoring, it is also possible to apply CP to pre-stressed bridges without risking hydrogen induced deterioration of the pre-stressed cables.

Austria also has several CP protected structures and currently the bridges along the Brenner freeway are being marked for renovation by CP.

5.2. Switzerland

The first CP installations in Switzerland were installed by Helbling Ingenieurunternehmung AG on a small bridge (surface approx. 30 m^2). Since then approximately 10 000 m^2 have been cathodically protected. The results have been mainly good. Problems have generally arisen only when the anodes were not properly laid (as to SGK Guidelines C7), the rectifier wrongly connected or an excessively high protective current injected. Three examples from the authors' experience which show the positive results of CP with concrete are given below.

5.3. Rodi Bridge

The bridge near Rodi on the Bellinzona-Airolo highway was opened for traffic in 1960. For almost 30 years, i.e. before the opening of the St Gotthard tunnel, there was heavy traffic including many heavy vehicles on this bridge; between 1964 and 1986 there was much use of road salt. For these reasons the bridge plates were renovated in 1984, although both the abutment to the right and left of the train tracks still contain the original construction material. The examination of 1988 showed that the concrete in the walls of the abutments within and without was heavily contaminated with salt (up to 2% chloride by weight of the cement; in certain places down to 60 mm depth). Potential measurements gave an alarming picture of heavy and very inhomogeneous corrosion on the rebars of the abutment walls. A visual judgement of the rebars showed that the cross-section reduction of the steel was not yet very

Fig. 1 *Preventive cathodic corrosion protection on bridges in Italy. The titanium net is fixed on the completed road slab and covered only by a layer of concrete.*

extreme. It was decided therefore to protect the existing structure by CP from further corrosion without removing the chloride-contaminated concrete.

On abutment I nets of titanium were installed at the outside of the post on 3 sides (Fig. 2). Abutment III was covered on the inner side with an anode net. The protected surface area of abutment I is 81 m², abutment III over 200 m². An unusual feature of this installation which is worth mentioning, is that the train tracks of the railroad under the bridge are at the same potential as the rebars. Stray currents from the CP system to the railroad tracks can be therefore excluded. On the other hand, tests have shown that the train currents (a.c.) are not a factor in the operating dependability of the CP system. Further, the earth anchor of abutment I is also connected to the rebars so that they too receive partial cathodic protection, although it is not a full scale protection.

Fig. 2 *Abutment I of Rodi Bridge in Canton Ticino. On 3 sides the anode net is installed.*

After installing the anode nets a protective current of about 15 mA per square metre of protected concrete surface was fed in. Tests made about one month after the opening of the CP system showed that even after such a short time a switch off potential of −820 to −890 mV vs CSE (copper/copper sulfate) could be measured. The depolarisation after 4 h was between 240 and 360 mV and thus easily fulfilled the minimum requirement of 100 mV.

Since activating the CP system in 1988 control tests have been made twice a year. On and off potentials, applied protective current and depolarising measurements are made. In addition, the voltage at the rectifier and the total delivered protective currents are measured and recorded by employees of the owner of the system every month. Table 1 gives an overview of the relevant measured values.

The measurements show that the switch off potential clearly fell shortly after the activation of the CP. This is due to migration of the chlorides away from the rebars which thereby encouraged passivation. It should be pointed out that the CP system was always adjusted according to the measurements to guarantee an optimal operational performance. The same is true for the interpretation of the protective current density.

At the time when this system was installed, the attachment of probes to control the current of protection had not yet become a standard technique.

As a result of the migration of chlorides and the repassivation of the rebars and because of the adjustment of the rectifier, the 100 mV criterion was not always obtained. Nevertheless, according to the values in Table 1 it can be concluded that collectively the rebars and both abutments are permanently protected from rust by the CP system although they are, as before, in excessively chloride-contaminated concrete.

Table 1. Control measurements CP Rodi Bridge

	Abutment I			Abutment III		
	I mAm^{-2} (concrete)	U_{off} V_{CSE}	Depol (mV)	I mAm^{-2} (concrete)	U_{off} V_{CSE}	Depol (mV)
Opening	14.81	−0.850	285	12.94	−0.870	350
1989	19.75	−0.545	175	14.93	−0.538	140
1989	20.02	−0.543	180	15.03	−0.476	175
1990	12.22	−0.493	145	8.86	−0.401	90
1991	13.21	−0.471	155	9.31	−0.433	85
1991	13.21	−0.430	160	14.39	−0.480	80
1992	13.09	−0.467	190	14.43	−0.480	100
1992	13.21	−0.530	180	12.14	−0.480	95
1993	14.07	−0.480	170	14.28	−0.471	95
1993	13.33	−0.467	150	10.20	−0.430	90
1994	14.32	−0.476	140	11.69	−0.416	80
1994	14.20	−0.470	120	11.29	−0.393	95
1995	13.46	−0.451	150	11.49	−0.410	80
1995	12.35	−0.464	130	9.85	−0.406	85
1996	12.04	−0.454	120	9.55	−0.412	90

5.4. Intschi Bridge

The Intschi bridge is a freeway bridge more than 40 m high over the river Reuss in Canton Uri (Fig. 3). An unsealed entrance into the hollow case meant that the right flank of the north post on the lane LORA (mountainwards) was covered for years with water of a high salt content. Due to the high chloride level and the resulting corrosion of the rebars the support safety of the post and their durability was in question. A static check later revealed that the post was at the limit of its load capacity. A large volume removal of the chloride-contaminated concrete was impossible without a long-term closure of the freeway. It was therefore decided to protect the

Fig. 3 View of the Intschi Bridge in Canton Uri. The post in the foreground was reconstructed.

posts from corrosion with CP. In order to improve the static strength additional reinforcement of the post with a reinforced concrete layer of 20 cm was planned. To protect the old chloride-damaged as well as the new rebars from corrosion and to keep the installation and CP control at a minimum, a new 'Sandwich-method', never previously tried for the CP was applied. The anode strips were attached to the standing concrete (Fig. 4) and finally the new reinforcement placed and cemented in. The problem with this method was to ensure that absolutely no electrical contact should arise between the old plus the new reinforcement and the anode. This construction method and the CP plan that was used is described in a separate article in the SIA journal [9].

To survey the CP in this structure sensors and macroelements were embedded and the wiring in both cases was executed in such a way that the protective current can be measured on a measuring device at the foot of the post. The potential values can be measured on permanently installed reference electrodes. The depolarisation

Fig. 4 *Anode strips on post of Intschi bridge. They were fixed by plastics holders.*

measurements can be read out by remote control with modem and computer. The results of these checks are represented on the graphs in Fig. 5.

It can be seen that with few exceptions the 100 mV criterion was met at each control reading and so the CP protection for the old as well as the new reinforcement is guaranteed. The reference electrodes are distributed over the entire structure and precisely attached to the heavily contaminated areas. So, according to the depolarisation values it can be confirmed that the entire protected post area is effectively protected against corrosion. The cyclic changes are due to the climatic influences (primarily temperature and concrete humidity.

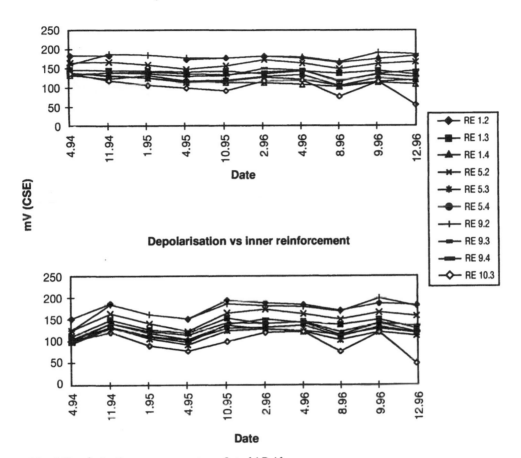

Fig. 5 *Depolarisation measurements on Intschi Bridge.*

5.5. Further, Recently Installed Systems

In addition to the above mentioned systems which have operated successfully for several years in Switzerland reinforced concrete constructions such as piers and supports, parking decks and tunnel walls, have increasingly been protected with CP in the past years. Tunnel walls were the largest surface area to be protected. On the National Freeway A2 in Canton Uri the Gotthard pre-tunnel (the small tunnel ~ 100 m in length acting as the tunnel entrance before the main tunnel) (Fig. 6) and in Canton Lucerne the parallel roadway of the Eich tunnel (Fig. 7) were cathodically protected. On the A5 in Canton Neuchâtel and — as this article is presently being written — the Tranchée couverte d'Auvernier (covered secton of Auvernier) is being renovated by the use of CP. The above mentioned projects were always treated in the splash water area of the tunnel walls. The work done on the Gotthard pretunnel [10] is of special technical interest and this application is therefore briefly mentioned here.

Fig. 6 *Gotthard Pre-tunnel. The posts and the walls were protected by CP.*

Fig. 7 *Eich Tunnel in Canton Lucerne. The wall of the parallel roadway was restored by CP.*

A parallel running emergency roadway is situated on both sides of both directions of the 398 m long Gotthard pre-tunnel and is separated from it by a 40 cm reinforced wall. The side of this wall on the driving area side was covered with chloride-containing water from the passing traffic. A high level of chloride developed in the concrete which caused corrosion (pitting). The side of the wall on the emergency roadway side showed signs of the onset of corrosion. The Gotthard pre-tunnel was restored in 1994/95 with CP.

The application of the CP was on the driving area direction side of the wall (Fig. 8). The goal was to protect both layers of rebars (directly behind the titanium net as well as the rebars 40 cm further distant) against further corrosion. Thirty-two reference electrodes and 23 current sensors (cut rebar crosses) were implanted as a survey system. Table 2 shows a survey of February 1996 values recorded only 6 months after the opening of the system.

It can be seen that for the front layer of rebars 11 of 12, and for the back layer 10 of 11 measuring sensors were already showing current flow and therefore an operating CP a few months after initiating the system. The measuring sensors are distributed over the entire structure and specifically attached to heavily contaminated areas and according to the confirmed protective current on the measuring sensors it can be concluded that the entire operating area is effectively protected against corrosion.

From the depolarisation values it is clearly seen that with only one exception the 100 mV criterion was satisfied within a few months of the opening using the reference electrodes at the front layer of rebars. The reaction of the back layer of rebars is of special interest. The depolarisation was between 30 and 167 mV and means that the back layer also received cathodic protection. The protection criterion was not met in all areas, but the rate of corrosion is clearly reduced as a result of the lowering of the

Fig. 8 Titanium anode net on the driving area direction side of the wall to the parallel roadway.

Table 2. Cathodic protection data from the Gotthard pre-tunnel 6 months after opening the system

Distance from portal (m)	Height from road (m)	Mountainside track				Valleyside track			
		Front		**Back**		**Front**		**Back**	
		Depol (mV)	I (mAm⁻²) (steel)	Depol (mV)	I (mAm⁻²) (steel)	Depol (mV)	I (mAm⁻²) (steel)	Depol (mV)	I (mAm²) (steel)
29	0.5	96	0.25	39	0.06	102	0.23	72	0.06
30	1.5	105	0.65	56	0.06	137	*	96	*
131	0.5	136	0.49	**	*	114	0.15	126	0.06
132	1.5	177	0.23	107	0.02	75	*	107	*
267	0.5	113	0.17	64	0.00	96	0.00	82	0.03
268	1.5	149	0.44	**	0.04	180	*	167	*
341	0.5	115	0.14	84	0.03	97	0.06	63	0.03
342	1.5	167	0.06	127	0.02	114	*	115	*

* No measuring sensors.
** Missing results.

potential. The result is that flow of a protective current through a reinforced, 40 cm thick concrete is possible, and that CP is the only method in which a one side application permits protection for both layers of rebars. In the gallery and tunnel walls which are subjected) to corrosion from the mountain side (e.g. aggressive mountain water, CP can be applied to the driving area side and the corrosion halted.

6. Conclusions

According to three examples from the authors' own experience and from a short general introduction and review of experience elsewhereit has been shown that CP today represents a procedure which should be considered as an alternative for restoring endangered structures in above and underground constructions. The advantages of this method (little concrete removal, total safety against corrosion, long life-span, possibility of monitoring efficiency etc.) must be weighed against the possible disadvantages (price, control time after renovation) and be compared to the conventional restoration methods. Even for new constructions which are exposed to severe conditions (mountains, ocean coast, structures with possibly contaminated water flow) the application of CP is advantageous.

References

1. R. F. Stratfull *et al.*, Transportation Research Record 1975, (539), 50–59.
2. F. Hunkeler, KKS — Wissensstand, Einsatzmöglichkeiten und Einsatzgrenzen, "CP state of the art, possibilities and limits". Swiss Society of Engineers and Architects, Documentation D02 1 , 27–42, (1988), Zurich.
3. Rule C7 of Swiss Society for corrosion prevention, Pfingstweidstr. 30, 8005 Zurich, 1991.
4. European Norm CEN TC262/SC2/WG2 (in preparation).
5. R. F. Stratfull, Criteria for the cathodic protection of bridge decks, in *Corrosion of Reinforcement in Concrete Construction*, Ed. A. P. Crane, Soc. Chem. Ind., London 1983, pp.286–331.
6. Internal report of Kenneth C. Clear Inc., 1985 (Eds W. E. Perenchio, J. R. Landgren, K. C. Clear and R. E. West), Cathodic Protection of Concrete Bridge Structures, Final Report NCHRP 12-19B, 1985.
7. D. M. Harriott *et al.*, Cathodic Protection of Reinforced Concrete Bridge Elements: A State-of-the-Art Report, SHRP-S-337 report, 1993.
8. US Departement of Transportation/Federal Highway Administration, Memorandum: FHWA Position on cathodic protection systems, 1982.
9. *Schweiz. Ing. Archit.*, SI+A, 10, 1997.
10. *Schweiz. Ing. Archit.*, SI+A, 30–31, 1995.

17
Cathodic Protection of Buried Reinforced Concrete Structures

Z. CHAUDHARY and J. R. CHADWICK

Taywood Engineering Ltd., 345 Ruislip Road, Southall, Middlesex, UK

ABSTRACT

Design constraints and considerations for the cathodic protection of buried reinforced concrete foundations are described. Negative potential shift and depolarisation results are described and discussed in relation to the 'Protection Criteria' for buried reinforced concrete structures. Adequate protection was achieved at all parts of the structures at an average applied steel current density of 15 mAm^{-2}. Depolarisation rates have been found to be significantly slow and vary from one part of the structure to another relative to the depths below grade level. The '100 mV instant-off shift' criterion is recommended for the performance assessment of CP systems for buried concrete structures.

1. Introduction

Cracking and deterioration of equipment foundations, in the process area of an operating oil refinery located near the Red Sea coast in Saudi Arabia, was observed at several locations throughout the plant, just seven years after it had been built. The majority of the equipment foundations are of a spread footing design with most of the concrete surface being below grade level at a depth ranging between 0.5 and 3 m. A concrete remediation project was initiated to determine the cause of the deterioration and to develop repair schemes.

Phase I investigations, completed in August 1992, concluded that deterioration of the concrete was due to chloride-induced corrosion of the reinforcement. The findings recommended that all critical structures should be repaired within the next six years using either structural repair or cathodic protection repair methods. During Phase II of the remediation project, a pilot CP repair scheme was designed, installed and monitored to develop a standard CP design approach for other foundations. Detailed design development and performance of the pilot CP system are described and discussed elsewhere [1]. In Phase III of the remediation project, CP systems for 15 foundations were designed and successfully commissioned. This paper describes the design considerations and performance of these CP systems during the first 12 months after commissioning and also discusses protection criteria for buried structures.

2. Design Constraints and Considerations
2.1. Anode Design

The designs had to overcome operational constraints which included a lack of access to the bottom face of the foundation slabs. An activated titanium mesh anode system was designed for the accessible concrete surfaces, i.e. top and side surfaces of the base slabs and associated columns, both for its ability to deliver a high level of current and for its durability under damp and saline environments. Preliminary studies had revealed that the bottom reinforcement of the base slab, particularly centrally, might not receive adequate and uniform protection from mesh installed on the top of the base slab, even if it were extended to the slab sides. Therefore, an independent anode system, comprising pre-packaged discrete high silicon iron chrome (HSI) anodes which have a good track record when used for the cathodic protection of buried pipelines, was designed to protect the bottom reinforcement (Fig. 1).

In view of the large width or diameter of some of the protected slabs (15–20 m), 'close ground beds' distributed around the slab were designed for HSI anode placement. This design approach was adopted to:

- Ensure sufficient protection to central areas of the protected slab which would be shielded from a remote anode.

- Minimise the current leakage to adjacent foundations, hence reducing the risk of stray current corrosion.

The design also ensured that the slab edges would be outside the 'area of influence' surrounding each ground bed and variations in applied current densities, between the nearest and the farthest points from ground bed, would not be in excess of 15%.
Steel density varied significantly at many locations in the top of the foundations.

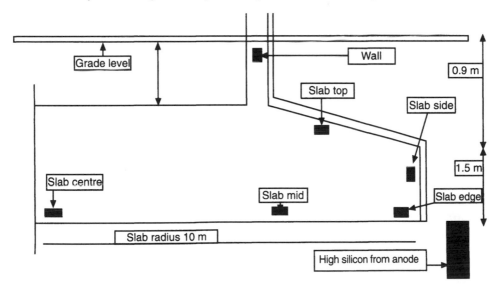

Fig. 1 *Typical sketch showing different parts of foundation slabs and CP monitoring positions.*

Protecting these areas in a common anode zone, could result in significant uneven current distribution across the top of the slab. On the other hand designing an independent anode zone for each area with different steel density would not be economical. This design constraint was overcome by varying the degree of expansion of mesh anode (i.e. using different grades of mesh anode) appropriately to match the variations of the steel density of the slab.

2.2. Design Life and Design Current Density

All CP systems with both consumable and non-consumable components, were designed to last for a life of 30 years without any major repair. The pilot CP system monitoring data [1] suggested that a steel current density of 15 mAm^{-2} was required to afford sufficient protection to all parts of the structure. Taking this into consideration plus an allowance for worst conditions, a steel current density of 20 mAm^{-2} was used to design the CP systems.

3. Monitoring Results
3.1. Potential Shift

The CP systems were energised at an average steel current density of 15 mAm^{-2}. Within 15 days after energisation, 100 mV potential shift was achieved at 39 (81%) monitoring locations out of the total of 48 (Fig. 2). At only 5 monitoring references, located at the bottom centre of the base slab, was it found to be <75 mV. The extent and rate of polarisation of steel potential in the negative direction was noticed to be relatively small and slow at the bottom centre and mid of the base slabs (Fig. 3). Nevertheless, the shift in negative direction was generally found to increase with time and 12 months after system operation, more than 100 mV had been achieved at 47 out of the total of 48 locations (Fig. 2). Slab edges relatively attained more shift due to receiving current both from the mesh and ground anodes. The magnitude of shift in areas with different steel densities was more or less balanced, indicating even current distribution in these areas.

3.2. Potential Decay

Initial performance verification of the CP systems was carried out 15 days after initial energisation. The potential decay rate was observed to be slower than that normally seen on atmospherically exposed structures (Figs 4 and 5); 100 mV decay (4 h after current interruption) was recorded at 8 out of the total 48 monitoring locations. The decay magnitude at each monitoring location, however, is found to increase with increase in time allowed for decay even 120 h after current interruption (Fig. 7). At 48 h after current interruption, 100 mV decay was observed at 23 and 26 locations after 15 days and 12 months respectively (Figs 4 and 5). There has been a gradual increase in decay rate over a period of 12 months.

Potential decay rate and magnitude also varied with location (Figs 6–8). Decay was slower and of a smaller magnitude at the centre of the base slab and higher in areas above or slightly below grade level (wall & slab top).

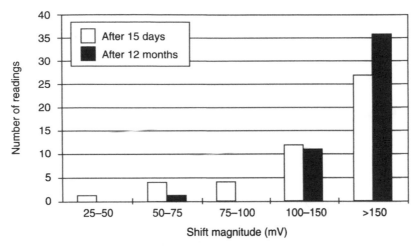

Fig. 2 *Changes in potential shift magnitude with time.*

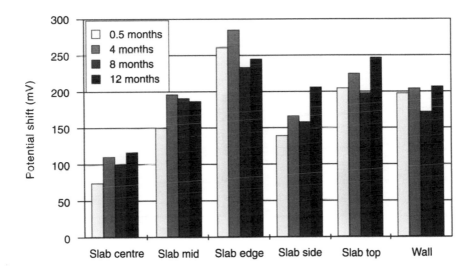

Fig. 3 *Average potential shift in different parts of the structure.*

4. Discussion

The achievement of 100 mV polarisation in the negative direction at almost all the monitoring reference points, suggests that the corrosion rate of the reinforcing steel in equipment foundations has been either significantly reduced or arrested. This implies that CP system for each structure is already meeting the design objectives within 12 months of its operation.

The lower rate and magnitude of potential shift observed at the bottom face of the base slabs, can be attributed to the associated anode system. Unlike the mesh anodes

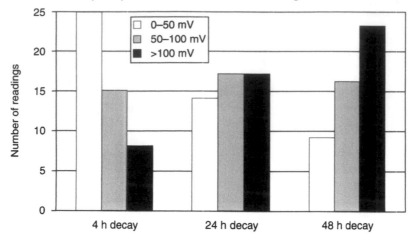

Fig. 4 *Potential decay magnitude — 15 days after commissioning.*

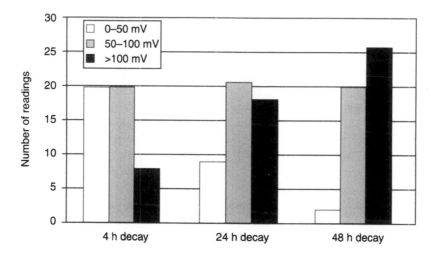

Fig. 5 *Potential decay magnitude — 12 months after commissioning.*

mounted on the concrete surface with a cementitious overlay, not all the current discharged by the silicon ground anodes is directly transferred or returned through the reinforcement to be protected. Some proportion of it is lost to adjacent foundations or pipework bonded to the protected structure. Nevertheless, results have indicated that inaccessible concrete surfaces of buried foundations can be sufficiently protected using conventional discrete high silicon iron anodes provided that they are appropriately designed.

The rate of depolarisation is generally found to decrease with increasing depth below grade level. This trend is consistent with that observed for the partly and fully submerged concrete structures by others [2,3]. This does not, however, suggest that protection to the bottom of the slab was inadequate, as steel potentials continued to

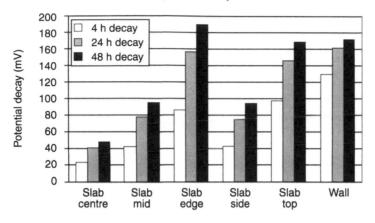

Fig. 6 *Average potential decay — 15 days after commissioning*

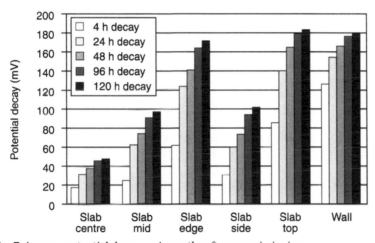

Fig. 7 *Average potential decay — 4 months after commissioning.*

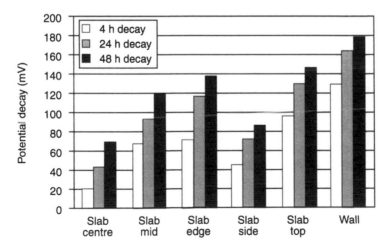

Fig. 8 *Average potential decay — 12 months after commissioning.*

decay towards less negative potentials when decay was allowed for longer periods. This observation together with the decay trend noticed for different depths below grade level, indicates that the lower rate of depolarisation is associated with slow ingress or diffusion of oxygen [4] to areas well below grade level.

The results suggest that 4 or 24 h depolarisation (potential decay) criterion, usually recommended for atmospherically exposed concrete structures [4,5], cannot be applied to partly or fully buried concrete structures. Depolarisation criterion is favoured [5] because it has the virtue of being independent of the IR drop error and also of the changes in rest potential as a result of increase in alkalinity due to cathodic protection. The resistivity of buried structures is usually significantly lower than that for atmospherically exposed concrete structures, therefore the IR drop error risk associated with the 'potential shift' criterion would be negligible and is in any case, eliminated when using the 'instant-off' technique. Therefore, we propose that the 100 mV shift (instant-off) rather than the decay is used for routine performance assessment of CP systems on buried structures. However, in order to determine the true polarisation, it is also recommended to establish new rest potentials after every six months, or at other suitable time intervals depending on the extent of changes in rest potential usually detectable from a significant decrease in shift magnitude, by allowing steel potentials to decay for longer periods.

5. Conclusions

1. Adequate protection of buried reinforced concrete foundations was achieved at an applied current density of 15 mAm^{-2} of steel.

2. High Silicon Iron ground anodes can provide adequate protection to inaccessible concrete surfaces of buried concrete structures provided that they are appropriately designed.

3. Depolarisation in buried foundations can be extremely slow depending on the depth below grade level as a result of slow ingress or diffusion of oxygen to such depths. This implies that potential decay criterion would be misleading in assessing the effectiveness of CP system for buried structures.

4. The 'instant-off shift' criterion can provide better assessment of CP systems for buried structures.

References

1. Z. U. Chaudhary, J. R. Chadwick and D. C. Pocock, The design and commissioning of cathodic protection system for buried reinforced concrete structures, in *Proc. 5th Int. Conf. on Deterioration and Repair of Reinforced Concrete*, Oct., 1997. The Bahrain Society of Engineers.
2. B. W. Cherry, Protection of buried and submerged structures, *Int. Conf. on Structural Improvement Through Corrosion Protection of Reinforced Concrete*, 1992, London. Organised by The Institute of Corrosion, Leighton Buzzard, Beds, UK.

3. G. K. Glass, Analysis of data on a reinforced concrete cathodic protection system, *Mater. Perf.* 1996, **35**, 2, 36–41.

4. Cathodic Protection of Reinforced Concrete, Concrete Society Technical Report No. 36, The Concrete Society, London, 1989.

5. BS 7361, "Cathodic Protection — Part 1. Code of Practice for land and marine applications".

List of Abbreviations

The following abbreviations occur in the text and in the Index of contents.

a.c.	alternating current
ASR	Alkali Silica Reaction
CE	Counter Electrode
CP	Cathodic Protection
CSE	Copper/Copper Sulfate Electrode
d.c.	direct current
ECE	Electrochemical Chloride Extraction
EIS	Electrochemical Impedance Spectroscopy
ER	Electrochemical Realkalisation
MFP	monofluoro phosphate
OPC	Ordinary Portland Cement
RC	Reinforced Concrete *or* Resistance Capacitance
RE	Reference Electrode
R.H.	Relative Humidity
SCC	Stress Corrosion Cracking
SSW	Synthetic Sea Water
ToF-SIMS	Time of Flight-Specific Ion Mass Spectroscopy
XPS	X-ray Photoelectron Spectroscopy
w/c	water/cement ratio
WE	Working Electrode

Index

Plate Section

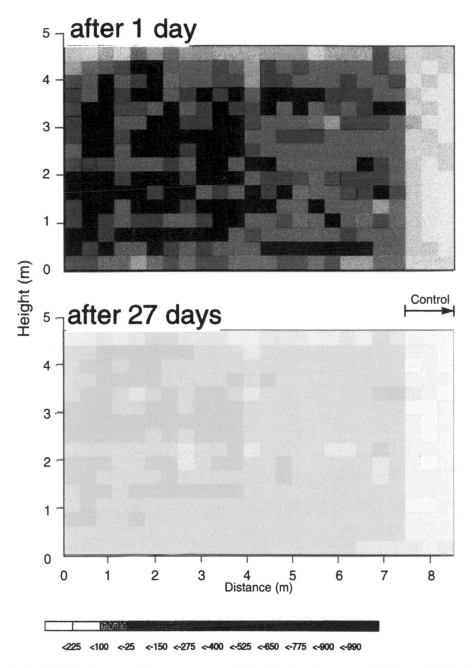

Fig. 8 *Half-cell potential measurements before (top) and one month after realkalisation treatment (bottom). Note that the three lines at the right correspond to the untreated control field.*

The above colour plate is from Chapter 11, page 136 of this volume.

T - #0521 - 071024 - C2 - 244/170/10 - PB - 9780367447687 - Gloss Lamination